RESTRICTED

FLIGHT MANUAL
B-24D AIRPLANE

THIS MANUAL IS CORRECT AS OF SEPTEMBER 15, 1942

THIS DOCUMENT CONTAINS INFORMATION AFFECTING THE NATIONAL DEFENSE OF THE UNITED STATES WITHIN THE MEANING OF THE ESPIONAGE ACT, U. S. C. 50:31 AND 32. ITS TRANSMISSION OR THE REVELATION OF ITS CONTENTS IN ANY MANNER TO ANY UNAUTHORIZED PERSON IS PROHIBITED BY LAW.

©2006-2009 Periscope Film LLC
All Rights Reserved
ISBN #978-1-4116-1321-8

COCKPIT CONTROLS
Used in Flying the B-24D

Unlock access door eye height, right side, ahead of bomb doors. Reach in, pull auxiliary bomb valve and open doors. Take one last look forward to see that the pitot head covers have been removed. As you enter the plane the fuel shut-off valves and auxiliary hydraulic power switches are located immediately overhead. To the left as you enter the flight deck are the fuel sight gauges. Valves on the bottom of the gauges select the tank. For correct reading, Inclinometer outboard of gauges should read zero.

Seat adjustment controls for:

See Page 5	Up and down	Inboard control lever—outboard side of seat
	Seat tilt	Middle control lever—outboard side of seat
	Back tilt	Outboard control lever—outboard side of seat
	Fore and aft	Control lever—front of seat in center

Rudder pedals are adjusted for length; locking latch on inboard side of pedal, released by pushing inboard with foot.

Brake pedals are operated by toe pressure. Parking lever is at Pilot's right hand on pedestal. Press pedals and lift lever to lock; press pedals to unlock.

Controls lock	Co-Pilot's left hand on pedestal—lift to lock. Securing strap stowed over Pilot's head.
Ignition and Battery	Co-Pilot's right hand—outboard forward.
Master Switch	Two main battery switches above them. A.C. power at Pilot's right hand on pedestal.
Booster Pump, Starter, Primer and Oil Dilution Switches	Co-Pilot's left hand—Instrument Panel.
Cowl Flap Switches	Co-Pilot's left hand on pedestal.
Prop Control Switches	Pilot's right hand on pedestal.
Intercooler Shutters	Center of pedestal.
Tab Control Knobs	Pilot's right hand on pedestal.
Landing Gear Control Lever	Pilot's right hand on pedestal.
Wing Flap Control Lever	Co-Pilot's left hand on pedestal.
De-icer and Anti-icer Controls	Co-Pilot's left hand—instrument panel.
Defroster	Defroster tubes under windshield right and left.
Heater Switch	Co-Pilot's right hand—aft of battery switches.
Ventilators	Outboard ends of Instrument Panel.

RESTRICTED

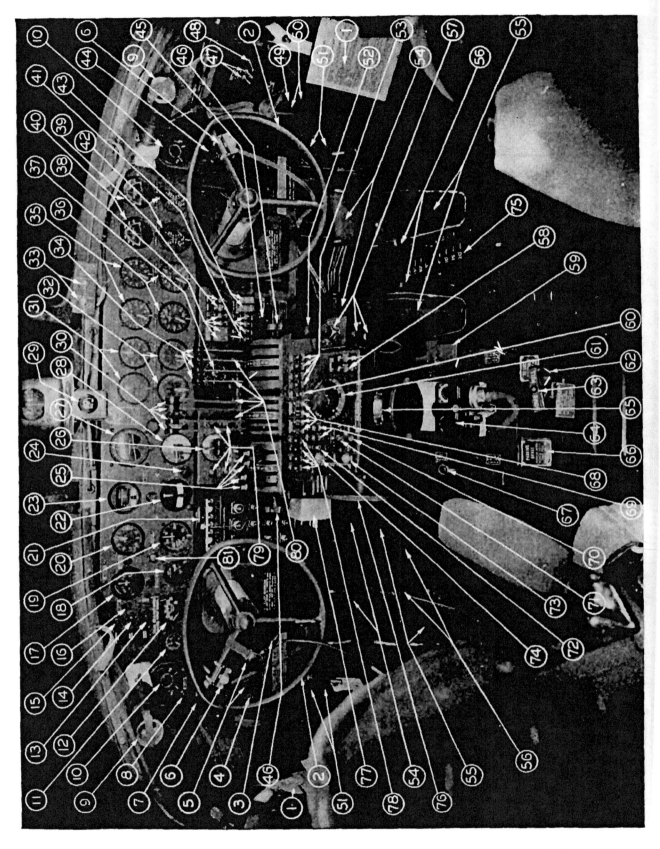

Page 2] COCKPIT CONTROLS

RESTRICTED

1. Flight Check-off List
2. Control Wheel
3. Push-Pull for Selection of Heat or Windshield Defrosting
4. Tell-Tale Switch
5. Hydraulic Pressure Gauges (Brakes)
6. Throat Microphone Switch
7. Hydraulic System Gauge
8. Rudder Tell-Tale Lights
9. Defrosting Tubes
10. Ventilator Control
11. Suction Gauge
12. Free Air Temperature Gauge
13. Camera Blinker Light
14. Bomb Release Light
15. Bomb Door Indicating Light
16. Tell-Tale Lights (4)
17. P.D.I.
18. Flap Indicator
19. Air Speed Indicator
20. Altimeter
21. A.F.C. Control Box
22. Turn and Bank Indicator
23. Turn Indicator
24. Marker Beacon
25. Supercharger Controls
26. Radio Compass
27. Gyro Horizon
28. Rate of Climb Indicator
29. Magneto Compass Light Rheostat
30. Throttles
31. Manifold Pressure Gauges
32. Tachometer
33. Destruction Switches
34. Mixture Controls
35. Fuel Pressure Gauges
36. Oil Pressure Gauges
37. Anti-Icer Rheostat
38. Booster Pump Switches
39. Primer Switches
40. Engine Cylinder Temperature Gauges
41. Starter Switches
42. Oil Temperature Gauges
43. Oil Dilution Switches
44. Formation Lights
45. Push-Pull for Selection of Heat or Windshield Anti-icing
46. Throttle, Mixture and Supercharger Tension Knobs
47. Battery Switches
48. Heater Switch
49. Master Magneto Switch
50. Magneto Switches
51. Oxygen Outlets
52. De-Icer Lever
53. Cowl Flap Switches
54. Brake Pedals
55. Rudder Pedals
56. Rudder Pedal Adjustments
57. Turn Control
58. Landing Light Switches
59. Wing Flap Handle
60. Pitot Heater Switch
61. Rudder Tab Knob
62. Controls Lock
63. Emergency Bomb Release
64. Radio Control
65. Aileron Tab Knob
66. Parking Brake
67. Landing Gear Handle
68. Intercooler Switches
69. Running Light Switches
70. Fluorescent Light Switch
71. A.C. Switches
72. Passing Light Switch
73. Alarm Bell
74. Horn Interruption Switch
75. Rudder Degree Markings
76. Elevator Tab Knob
77. Recognition Light Switch Box
78. Propeller Switches
79. Propeller Governor Limit Lights (4)
80. Fast Feather Circuit Breakers (4)
81. "Wheels Locked Down" Indicator

RESTRICTED

COCKPIT CONTROLS [Page 3

Landing Lights	Co-Pilot's left hand on pedestal.
Navigation and Formation Lights	Pilot's right hand on pedestal.
Recognition Lights	Pilot's right hand—left side of pedestal.
Fluorescent Panel Light	Pilot's right hand on pedestal.
Oxygen Regulators	Pilot's left knee and Co-Pilot's right knee.
Radio Interphone Control Box	Pilot's left elbow, Co-Pilot's right. Throat microphone button on wheel.
Mixture Controls	Right unit of control quadrant.
(Positions listed from forward to aft)	**Position 1. "Idle Cut Off:"** Stops engines by shutting off gas flow.
	Position 2. "Auto Lean:" Cruising position only. On the lean side. Not to be used at over 1400 RPM on run-up nor on operations requiring over 29.5 Hg. manifold pressure.
	Position 3. "Auto Rich:" Run-up; take-off; landing. Operations requiring more than 28 Hg. manifold pressure up to 14,500 feet altitude; over 14,500, 29.5 Hg.
	Position 4. "Full Rich:" Emergency, Cuts out automatic mixture control and sets carburetor to full power mixture requirements.
Throttle Controls	Center unit of control quadrant. "CLOSED" to "OPEN"
Turbo Supercharger Controls	Left unit of control quadrant. "OFF"—to—"ON"
A.F.C.E. Flight Control Unit	Turned on and handled by control switches on the A.F.C.E. panel at right of Pilot's control wheel axle.

For location of **ALL** controls used in flight, see "Location of Controls" section, Page 94.

29. Magnetic Compass and Light Rheostat
33. Destruction Buttons
79. Fast Feathering Switches

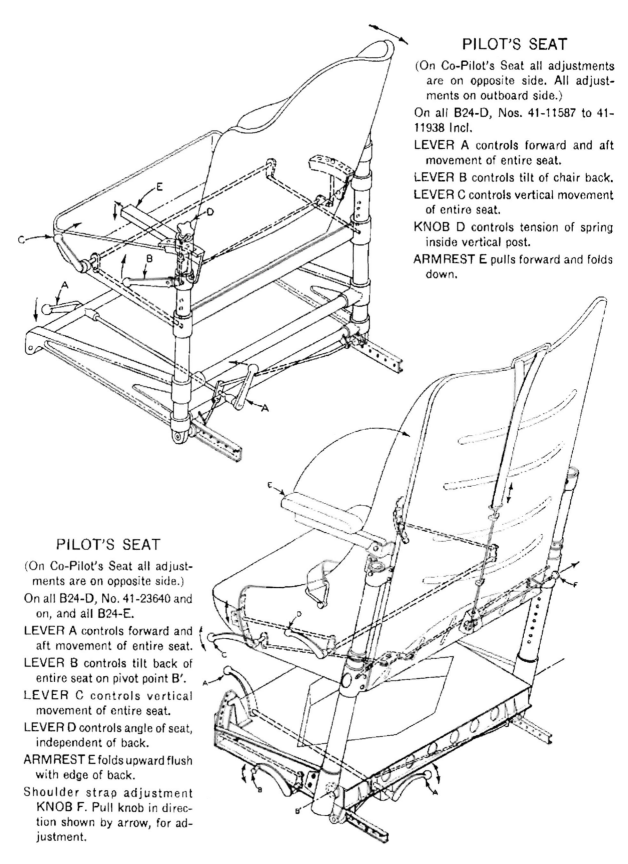

PILOT'S SEAT

(On Co-Pilot's Seat all adjustments are on opposite side. All adjustments on outboard side.)

On all B24-D, Nos. 41-11587 to 41-11938 Incl.

LEVER A controls forward and aft movement of entire seat.

LEVER B controls tilt of chair back.

LEVER C controls vertical movement of entire seat.

KNOB D controls tension of spring inside vertical post.

ARMREST E pulls forward and folds down.

PILOT'S SEAT

(On Co-Pilot's Seat all adjustments are on opposite side.)

On all B24-D, No. 41-23640 and on, and all B24-E.

LEVER A controls forward and aft movement of entire seat.

LEVER B controls tilt back of entire seat on pivot point B'.

LEVER C controls vertical movement of entire seat.

LEVER D controls angle of seat, independent of back.

ARMREST E folds upward flush with edge of back.

Shoulder strap adjustment KNOB F. Pull knob in direction shown by arrow, for adjustment.

RESTRICTED

COCKPIT CONTROLS [Page 5

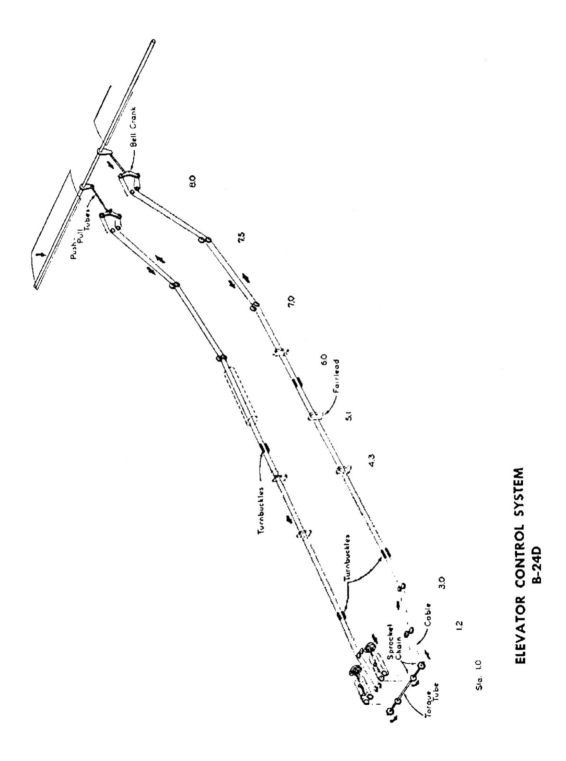

ELEVATOR CONTROL SYSTEM B-24D

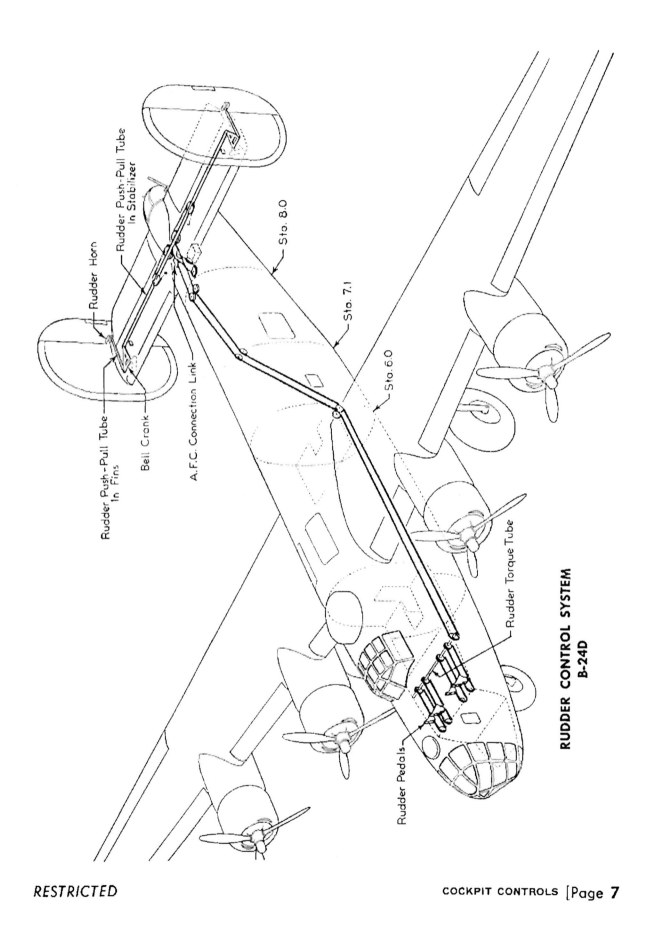

RUDDER CONTROL SYSTEM
B-24D

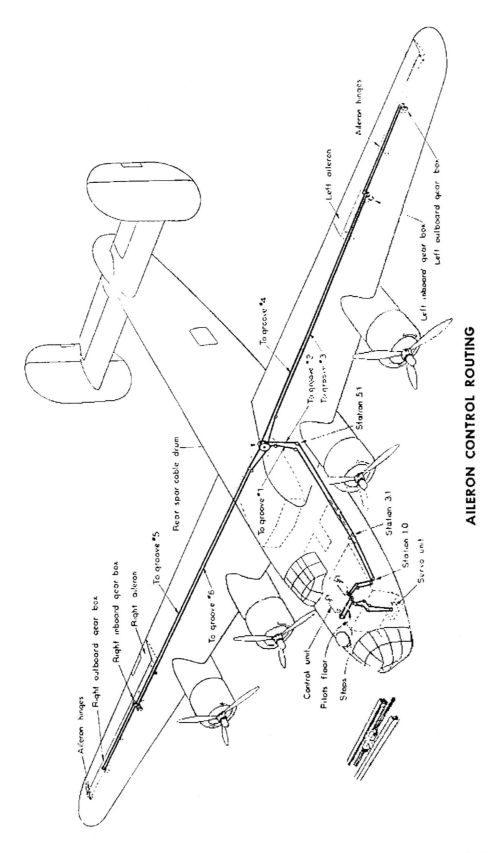

AILERON CONTROL ROUTING

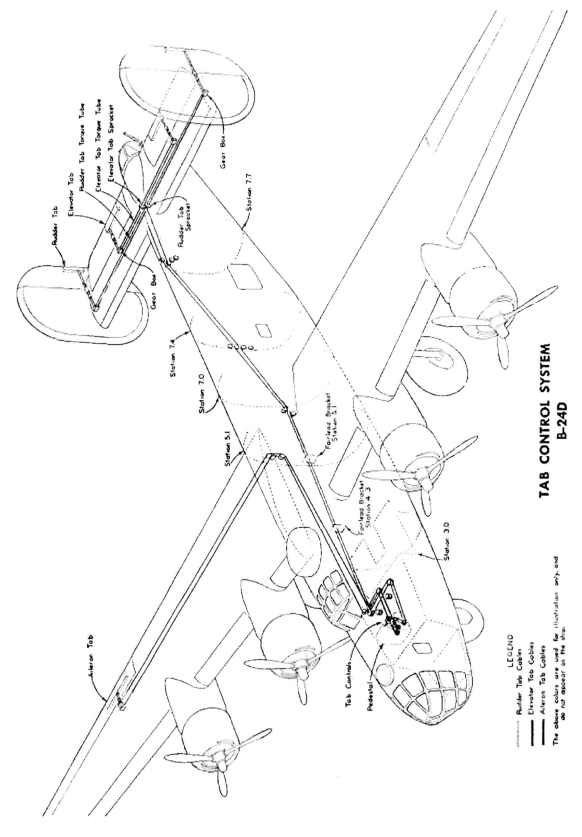

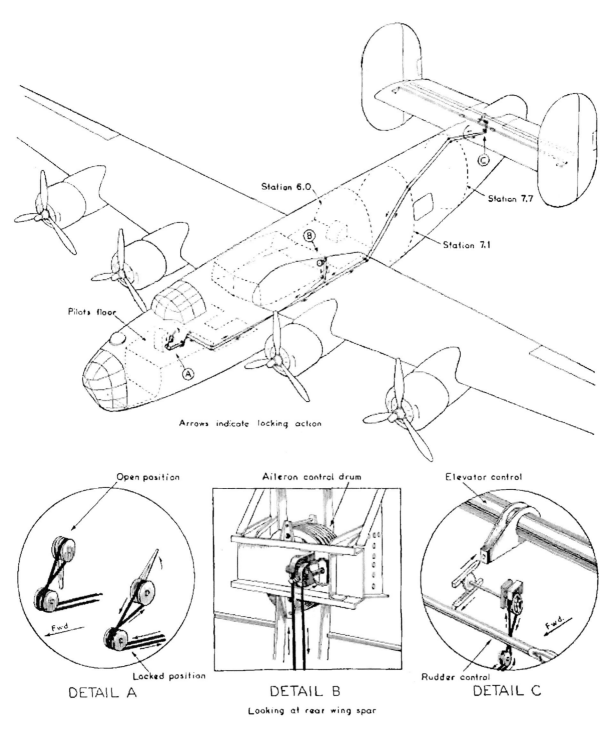

**CONTROLS—LOCK SYSTEM
B-24D**

POWER PLANT CONTROL SYSTEM
B-24D

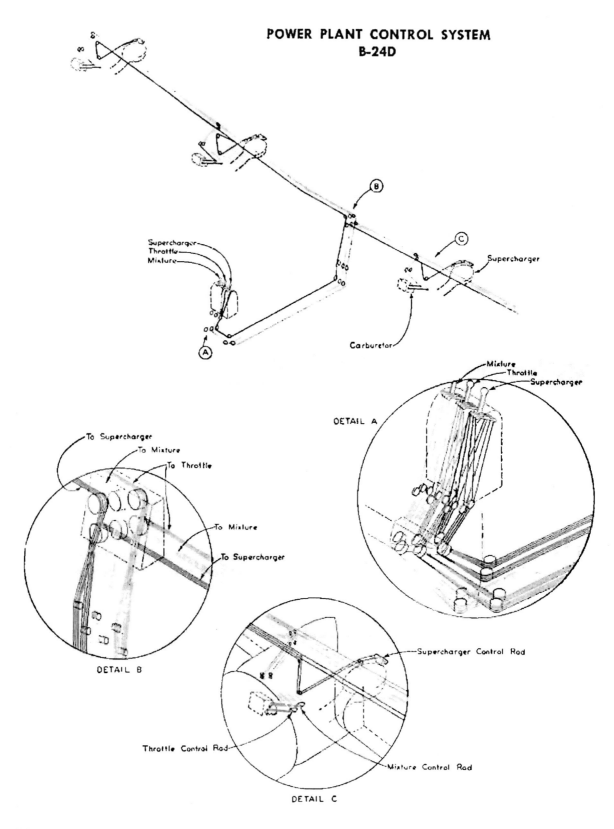

COCKPIT CONTROLS Page 11

WHY DRAG YOUR ANCHOR WHEN YOU CAN FLY!

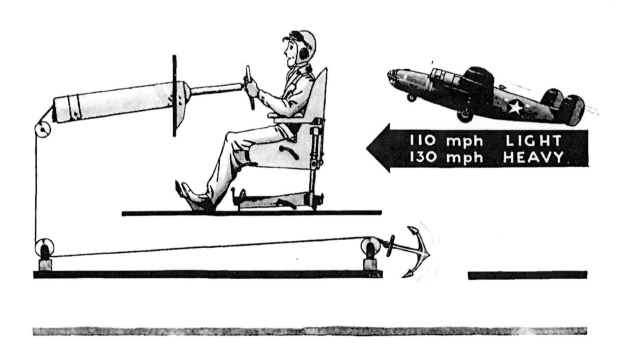

How to Fly

THE B-24D AIRPLANE

First and of foremost importance, you are the Pilot; the lives of your crew and successful completion of your mission is in your hands. Use good judgment and common sense. The airplane is a piece of machinery and will react exactly as you direct. It will not fight back nor argue with you, so do not get mad at it, it only affects your own reactions and corresponding ability to fly.

The following text on flying the B-24 airplane is based on experiences of Consolidated Pilots with many thousands of hours flying time in cooperation with officials of the U. S. Army Air Corps with their wide experience in flight training procedure.

The B-24 airplane is not difficult to fly. It has no vicious characteristics and when the Pilot learns the difference in "feel," due to its size, weight, and speed range, flying it is no more of a problem than flying a trainer. A large bomber is a highly complicated piece of equipment containing many compartments. Learn your airplane; study the functional operations of the several systems and the mystery of imagined complexities will become surprisingly simple. A little time devoted to the fundamentals of what makes it "tick" will pay amazing dividends in psychological reaction and peace of mind.

Master the airplane, don't let it master you, but—never lose respect for it.

Inspect your airplane before take-off or be sure competent hands have accomplished the pre-flight check. The importance of following the Pilot's Check-Off List carefully before every flight cannot be overemphasized. This check-off list specifies all the details that must be covered to insure safe flight. (See Page 27.)

When engines reach the proper operating temperature use discretion in taxiing away from the line. Sharp turns should be avoided. Sharp turns grind off rubber and apply serious stresses to nose and main gear. Use the engines for steering, and save the brakes. Taxi slowly; it is a simple matter to keep full control of the airplane with the engines.

Before take-off, run up engines, check magnetos, propeller control, and set superchargers in accordance with Pilot's Check-Off List. To avoid fouling plugs, idle engines at 800 to 1000 RPM. Close cowl flaps to one-third open. Extend wing flaps one-fourth for best normal average; one-half for shortest take-off. Head into wind, open throttles slowly and full out to stops. Hold brakes on until manifold pressure reaches 25" Hg. Use full power for take-off; it lessens the take-off run and corresponding wear and tear on landing gear, tires, and entire airplane. Have Co-Pilot hold throttle against stops and adjust superchargers so as not to exceed manifold pressure operating limits (consult operating chart before take-off to determine this). On take-off, maintain straight course with rudders. Do not use brakes.

CHECK DOWN LATCHES BEFORE YOU COME DOWN

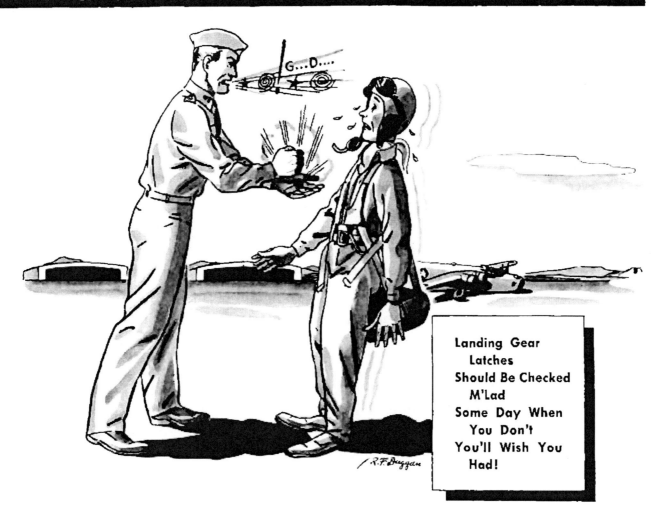

Landing Gear Latches Should Be Checked M'Lad Some Day When You Don't You'll Wish You Had!

As the plane picks up speed, lighten the nose and help it take-off at a safe minimum speed; this is 110 MPH at 45,000 pounds to 130 MPH at full load. Do not hold the airplane on the ground when it is ready to fly. Any idea of picking up extra speed on the ground as a safety margin is quite the reverse. Help your airplane.

Raise landing gear when definitely clear and air borne, and reach 130 MPH airspeed as soon as possible to be in best condition in the event of engine failure. (For engine failure see Page 92.) After take-off, maintain airspeed under 150 MPH until flaps are raised. The best average climb is 150 MPH. Consult performance chart for exact figure for specific load conditions. As long as the required minimum airspeed for stall is exceeded, the airplane is fully maneuverable with the flaps in any position. Stalling speed varies with loading, landing gear, cowl flaps, and flap setting. Maintain engine head temperature within limits given under "Power Plant," Page 31, by control of cowl flap opening. The cowl flaps cause buffeting between the one-third and two-thirds open position which should, therefore, be avoided.

If anything other than an airport flight is to be made, turn off auxiliary hydraulic master switch. See "Check-Off List" Page 27.

After reaching cruising altitude, level off, get "ON THE STEP" and pick up speed before power is reduced to cruising requirements. If power is reduced too soon and before the airplane has picked up full momentum for cruising it would mush along in a high attack, high drag attitude in trying to gain speed under reduced power and would probably be quite sluggish. Approach the cruising condition from the top, both speed and altitude, NEVER FROM BELOW.

The air handling of the airplane is conventional and normal. Stability is excellent and high maneuverability is possible. Primary instruction in flying has made the Pilot aware of load factors. Keep this in mind when banking or maneuvering so as to not exceed the safe limits. (See "Special Instructions" Page 25, for flying limitations.)

When you have squared away on a mission, check wheels, flaps and cowl flaps. Check the fuel supply frequently lest an unexpected leak or excessive consumption place you in a difficult position.

Two inches of boost gain by use of turbos is recommended as the best operation. (See "Turbo Supercharger," Page 48.) Too much boost will lean the mixture, evidenced by rise of head temperature. Too little boost will enrich the mixture with resulting loss of power and excessive fuel consumption.

Before entering the airport area, accomplish Pilot's Check-Off List, Page 27. Allow ample time to slow down to 150 MPH; turn on auxiliary hydraulic power; turn on booster pumps; lower landing gear and check the latches before making the final turn to enter the landing lane. Turbo superchargers "OFF" as the waste gate closes with reduced exhaust pressure when engines are throttled back. On entering the final landing lane slow to 140 MPH; extend flaps one-half. Extended flaps not only increase lift and drag but change the glide angle and attitude of the airplane in a manner to greatly increase visibility. Speed to be maintained in a glide varies, depending upon load, flap setting, and use of power. Under 45,000 lbs. loading glide should be maintained at 120 MPH slowing to 110 MPH with full flaps on leveling off for landing.

The airplane is fully maneuverable with flaps extended. Maintain sufficient RPM to con-

LOW BANKS AIN'T HEALTHY

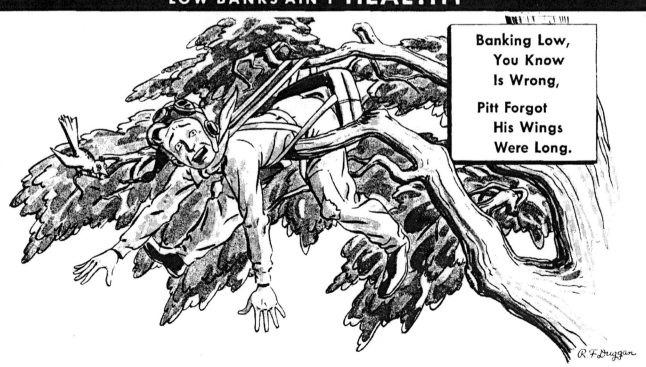

Banking Low,
You Know
Is Wrong,

Pitt Forgot
His Wings
Were Long.

tinue at a rate of descent of 400-600 feet per minute. At any time during the glide (but allowing ample time before crossing boundary of field to adjust to change of attitude before final stage of landing) extend the flaps fully. After crossing the boundary and over runway, close throttles fully and have Co-Pilot hold them to prevent creeping. As the plane begins to settle, hold it off the ground as long as possible. The exception to this is an emergency when it is necessary to use brakes immediately on touching the ground, which, too, is the only excuse for a three wheel landing. Hold the nose wheel off as long as possible and let the nose of the plane settle slowly and without shock, onto the nose gear. Do not "slap" the nose forward nor allow it to do so and do not apply the brakes with the nose wheel clear of the ground. (Crew aft will facilitate keeping the tail down, but do not exceed allowable C.G. limits for landing.) In case of emergency or of faulty brakes, a nose high landing with tail skid dragging (retractable type only) will enable the Pilot to land on any normal airport without using brakes.

Open cowl flaps immediately after landing and raise the wing flaps when convenient, but preferably before taxiing to avoid possibility of rocks or mud being thrown into the tracks.

Again taxi slowly and steer with the engines. Use brakes only when absolutely necessary. Enter parking area carefully. The wing span is 110 feet. There is no excuse for the carelessness of a collision on the ground or ground crash. Stop the engines with the mixture cut off. Leave cowl flaps open until engines cool. Set landing gear lever in the "DOWN" position. after No. 3 Engine has stopped. Do not set parking brakes until brake drums have cooled.

On parking the airplane, align the nose wheel to coincide with the center line of the airplane.

See **Page 92** for procedure for engine failure on take-off.

FLYING CHARACTERISTICS

General—Steep banks up to 60° can be made easily and safely. However, it should be borne in mind that in a normal bank of 60° the load factor is "2" and in this position all loads are twice as severe as in level flight. Turns steeper than normal increase this load factor.

Rough air operation is not critical. However, it is good practice to slow down to 150 MPH (240 KmPH; 130 Knots) when in extremely turbulent air, and extend the landing gear if flying on instruments. Disengage the Automatic Pilot when flying in rough air.

Longitudinal Stability—The longitudinal stability of the airplane is positive over a wide range of center gravity locations. Under normal loadings the airplane will return to normal flight when released from a stall or other abnormal positions. The maximum forward location of the C.G. should not exceed 23% of the mean aerodynamic chord while maximum aft location of the C.G. should not exceed 35% M.A.C. Care should be exercised to operate controls smoothly when flying close to these limits, especially with the C.G. in extreme aft positions, as it is easily possible to develop the limit load factors of the tail assembly with sudden heavy elevator operation.

In the higher speed range, the elevators become "heavy." This is desirable inasmuch as it helps to prevent sudden extreme application of the elevators, which might prove damaging to the structure. When maneuvering the airplane, as in a dive, always keep the airplane trimmed by use of the trim tabs. If the Pilot attempts to hold the full stick load, his sudden relaxing can apply a destructive force to the airplane.

Brakes and Taxiing—The brakes are operated by two sets of interconnected pedals from either the Pilot's or Co-Pilot's side. The hydraulic brakes operate on both an inboard and outboard drum on each main wheel. Failure of either the inboard or outboard system will leave ½ braking power. The operation of the hydraulic brakes is smooth and not "touchy." Increasing pedal pressure increases braking pressure proportionately.

The nose wheel is free to swivel 45° each way and is damped against shimmying. Any shimmy should not be tolerated for it can be cured by proper servicing of the dampers. Turning too sharply will damage the nose gear.

The airplane can and should be taxied by using outboard motors without the brakes.

Avoid overheating of brakes caused by applying them for long periods. Do not make small radius turns. Do not lock the inside wheel as it tears the rubber. Allow the inside wheel to roll. When necessary to use the brakes, they should be coordinated with applications of power.

The airplane has no inherent tendency to ground-loop and can be turned to either side while taxiing at a fast rate.

If the brakes have been used to any great extent prior to taxiing up to the line, allow the brakes to cool before applying the parking brake.

The main landing gear is located at approximately 40% of the M.A.C. C.G. locations forward of this point obviously will have no tendency to rock the airplane so as to lift the nose wheel off the ground, while movement of C.G. aft will prevent nose wheel contact.

Take-off—Take-off procedure is consistent with that of other large airplanes of the tricycle landing gear type and the ship will come off the ground easily at 110 MPH (175 KmPH; 95 Knots) for gross weights up to approximately 45,000 pounds (20,412 Kg.) to 130 MPH (210 KmPH; 115 Knots) for heavier loads.

After opening the throttle hold the brakes until the manifold pressure reaches 25" Hg. This permits the turbo regulator to stabilize and results in a smooth flow of power from all engines, and makes it much easier to maintain a straight course on take-off. (See "Turbo Supercharger.")

Climb—The most practical speed for the best average rate of climb is 150 MPH (240 KmPH; 130 Knots), 2550 RPM—41" to 45½" Hg.)--See "Climb Chart" for accurate data.

Stall—The stalling characteristics of the airplane depend, in addition to the inherent design, on wing flap setting, cowl flap position, landing gear position, the power setting, and whether de-icer system is operating.

The stalling point of the B-24D and B-24E airplanes is clean and forewarned by a tail shake and slight pitching. The complete stall is followed by the airplane falling off to either side without a tendency to spin. Due to the aerodynamic cleanliness of the design, the airplane will pick up speed rapidly. Correct the slight yawing produced by falling to one side or the other by application of rudder, **the use of ailerons to lift a wing under this condition is forbidden until flying speed is recovered.**

Extended wing flaps will reduce the stalling speed, as will also the use of power. Cowl flaps should be closed prior to stalling to avoid tail buffeting.

DON'TS

DON'T fail to carry out all items of check-off list.

DON'T fail to check fuel before take-off and at regular intervals during flight.

DON'T fail to have nose wheel accumulator checked before take-off.

DON'T forget to check the de-icers "OFF" before take-off or landing.

DON'T forget to check the AFC or Gyro Pilot "OFF" before take-off or landing.

DON'T start engines without first pulling propellers through six blades.

DON'T use starter for direct starting. Inertia wheel must be energized before meshing.

DON'T use ship's batteries for first starts when battery cart is available.

DON'T fail to use auxiliary power plant when using ship's batteries for all starts.

DON'T attempt to start engines with low batteries.

DON'T fail to turn auxiliary hydraulic electric pump "OFF" after take-off.

DON'T fail to check the landing gear latches "engaged" before landing.

DON'T use brakes for steering in taxiing.

DON'T turn too sharply. It will damage landing gear and tear tires.

DON'T transfer fuel while using Radio.

DON'T forget high fuel consumption—Shooting Landings 300 G.P.H.

DON'T fail to study Landing Gear Emergency System.

DON'T fail to study use of Intercoolers.

DON'T attempt to use intermediate positions on mixture control.

New Inertia starters are not direct drive—they are "Constant Energizing" and must be brought up to speed before "Meshing"/"Cranking."

DID YOU CHECK YOUR FUEL SUPPLY?

HOT LANDINGS HURT!

The loss of control in a stall is gradual, with all controls losing effectiveness at about the same time. With the de-icer boots operating, the stall is sharper.

With full flaps, use of a little power will reduce the stall speed appreciably.

Spins—The airplane has no inherent tendency to spin from a stall or slow, steeply banked turns and should not be forced to do so under any condition. The airplane was not designed for the loads imposed on the structure during a spin condition, and structural failure could result from spins.

Dive—The diving speed limits for various gross weights are:

	MPH	KmPH	Knots
41,000 lbs. (18,144 Kg.)	355	570	305
47,174 lbs. (21,398 Kg.)	325	520	280
56,000 lbs. (25,401 Kg.)	275	440	240

Air loads build up rapidly on any large airplane in a dive, therefore, avoid abrupt movements of the controls.

Control trim should be maintained with the idea of keeping tail surface forces at a minimum. **It is better to trim the airplane to slightly nose heavy rather than tail heavy.** If it were trimmed tail heavy, in a dive the inherent tendency to pull up would make the application of up-elevator easier and more abrupt, creating higher load factors of "g's."

Landing—A three wheel landing is neither desired nor necessary. Land tail down with the main gear touching first and as speed diminishes allow the plane to settle gently on the nose gear. Landings from power glides with 400-500 ft./min. (122-152 meters/min.) rate of descent are easiest to accomplish because of the difficulty of making the proper landing flare with higher rates of descent, especially with a heavy load.

By using power in the glide with full flaps, indicated speeds as low as 105—115 MPH (170—185 KmPH; 90—100 Knots) can be held without danger, high loads requiring the higher speeds. These low speeds are **not recommended** but are mentioned to give the **minimum** limits of operation.

With no power, the desirable indicated gliding speed is 120 MPH (190 KmPH; 105 Knots). This speed also permits 10° banks near the ground and allows sufficient maneuverability with flaps extended for landing in bad visibility.

It is desirable to have the cowl flaps closed during extended reduced power glides in order to prevent rapid cooling of the engines. Always close flaps for landing.

When flying a heavy airplane remember that the inertia of a heavy body in motion resists effort to change the direction of that motion. Therefore, if a steep glide is being made with consequent high rate of descent, it takes some time and a considerable force to flare out this rate of descent and change the direction to one parallel to the ground. It cannot reasonably be expected with a rate of descent of over 500 ft./min. (152 meters/min.) to start to level off 5 to 10 feet (1.5 to 3 meters) above the ground and succeed in doing anything but "flying in."

While the tricycle landing gear does permit certain liberties of landing technique it does not permit ground contact while still in the glide as has been accomplished with lighter airplanes.

Brake Pedals Aren't Footrests When Landing the Ship.
Ask "Cocky," He Knows, He Tried It Last Trip.

BRAKE IT EASY!

Normally the main landing gear wheels should touch the ground first, in the same manner as the tail wheel airplane. Brakes should not be applied until the nose wheel is on the ground and the weight of the plane is taken by the oleo strut. The airplane will tend to rock forward onto the nose as it loses speed and it should be prevented from doing so as long as possible with the elevators. If a sudden application of the brakes is made with the nose wheel off the ground, the tail will snap up and excessive load factors will be built up in the nose wheel gear and in the tail, due to the length of the airplane.

CAUTION: NEVER LAND WITH BRAKES LOCKED.

The emergency tail wheel or skid is not designed for landing loads and is only to protect the vertical fin and lower gun turret against accidental rocking back.

Airplanes equipped with a retractable tail skid may be landed with the tail low, as are airplanes without tricycle landing gear. The retractable mechanism, while not recommended for full tail landing loads, will stand the load imposed by rocking back after landing, and by "dragging" the skid. In case of brake failure this feature may be used to advantage particularly if the crew is stationed well aft in the tail to keep the skid on the ground during the full run.

In case of emergency where the shortest possible landing run is imperative the use of brakes immediately on landing is necessary. In this case a three wheel landing is made and the thrust of the nose gear must be taken up by pushing the elevator controls forward before applying brakes. Do not lock wheels as tires will tear through the fabric in an astonishingly short time.

The airplane has no tendency to ground loop in a cross-wind but any drift should be taken out before making ground contact.

Two Engine Failure—Even with two engines inoperative on one side, it is possible to fly the airplane in all normal maneuvers within the engine power limits. The use of full rudder tab greatly relieves the rudder pedal pressure required to maintain straight flight. When two engines on one side are delivering rated power it is more desirable to bank the airplane with the dead engines high. (See Page **92** for engine failure during take-off.)

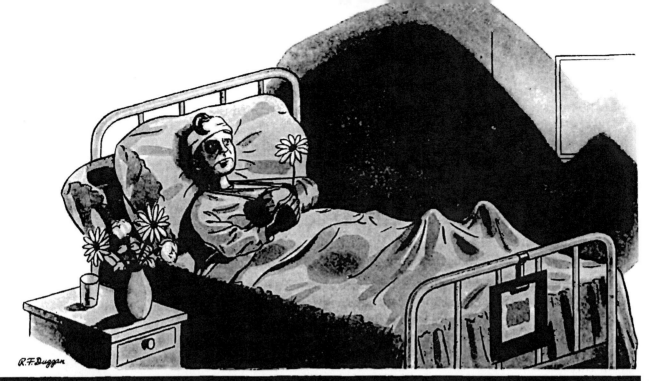

DIVING AIN'T HEALTHY

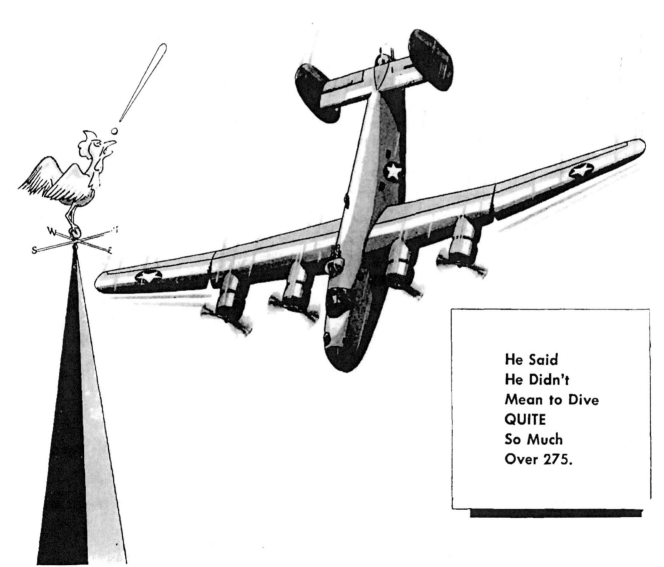

He Said
He Didn't
Mean to Dive
QUITE
So Much
Over 275.

SPECIAL INSTRUCTIONS

FLYING LIMITATIONS

Maneuvers Prohibited

    Loop   Stall
    Spin   Inverted Flight
    Roll   Dive (except as shown on Page 21)
    Immelmann   Vertical Bank

Only conventional flying is permitted when airplane is loaded to maximum loaded weight for safe flight.

Airspeed Restrictions

1. **Do not** exceed the following indicated airspeeds:

Gross Weight	MPH	KmPH	Knots
41,000 lbs. (18,597 Kg.)	355	570	305
41,174 lbs.	325	520	280
56,000 lbs. (25,401 Kg.)	275	440	240

2. **Do not** lower landing gear hydraulically at a speed in excess of 155 MPH (250 KmPH; 135 Knots).

3. **Do not** lower wing flaps or fly airplane with wing flaps down at a speed in excess of 140 MPH (225 KmPH; 120 Knots).

4. **Do not** operate de-icer system at take-off or landing.

Instrument Dial Limitation Markings

Airspeed and engine instruments are marked to indicate their operating limits as follows
    Green indicates the operating range.
    Red indicates the maximum and minimum safe reading.

Power Plant Restrictions

1. **Do not** exceed an engine speed of 2700 RPM.

2. **Do not** idle engines below 800 RPM for long periods.

3. **Do not** fail to operate oil dilution system for 3 or 4 minutes prior to shutting off engine, when frigid temperatures are anticipated.

Instrument Vacuum Selector Valve

In case of failure of either No. 1 or No. 2 Engine or of the air pumps on those engines, switch the vacuum selector valve on forward face of bulkhead at Station 4.1 so that the useful pump will operate the instrument suction rather than the de-icer suction.

Restrictions on Use of Automatic Pilot—or A.F.C.

Since experience with Automatic Pilots has demonstrated that abrupt control responses of the automatic mechanism under conditions of side slip or stall may result in a spin, the following must not be forgotten:

1. **Do not** operate airplane by the Automatic Pilot in extremely turbulent air under the following conditions:
   a. When de-icer system is working; or
   b. When one or more engines are not delivering normal power output.
2. **Do not** place airplane under control of the Automatic Pilot regardless of speed or altitude, until the Pilot has determined by manual operation that the existing conditions permit safe control by the Automatic Pilot. In no case will Automatic Pilots be used when the airplane is flying at less than an indicated air speed of 155 MPH (250 KmPH; 135 Knots).
3. **Do not** operate airplane under control of the Automatic Pilot without one rated Pilot remaining "on watch" and maintaining a close check of the airplane and instruments.
4. **Do not** engage Automatic Pilot when follow-up indices are not lined up.
5. **Do not** make course and altitude changes with rapid knob movements. Turn slowly and smoothly.
6. **Do not** allow airplane to get too far out of trim.
7. **Do not** turn any of the three speed controls to "OFF" or "LOWEST SPEED" when Automatic Pilot is engaged as this would lock the corresponding surface controls in whatever position they happen to be.
8. **Do not** forget that an Automatic Pilot can be overpowered.
9. **Do not** forget that an Automatic Pilot can "spill" and "wander."

Bomb Clearance Instructions

When releasing bombs in a glide or climb, observe the restrictions shown in "Armament Manual."

PILOT'S CHECK-OFF LIST

B-24D and B-24E Airplanes

BEFORE STARTING ENGINES:

1. Check Form 1 and Loading
2. Pitot Heads—Covers Removed
3. Wheel Chocks in Place
4. Bomb Doors and Cabin Doors—"OPEN"
5. Fuel Tank Valves—"ON"
6. Amount of Fuel—not less than 1200 U. S. Gallons
7. Main Line and Battery Switches—"ON" (If Battery Cart not used)
8. Generator Switches—"OFF"
9. Auxiliary Power Unit Started
10. Turn on Electric Auxiliary Hydraulic Pump
11. Adjust Seat and Rudders
12. Parking Brake—"ON"
13. Instrument Power Switch—"ON"
14. Navigation and Cabin Light—"ON" (Night)
15. Supercharger—"OFF"
16. Mixtures—"IDLE CUT OFF"
17. Automatic Pilot—"OFF"
18. Wing and Prop De-Icers—"OFF"
19. Intercooler Shutters—"OPEN"
20. Cowl Flaps—"OPEN"
21. Altimeters Set
22. See that Propellers are Clear and Ground Crew Notified, Fire Guard Posted, Pull Propeller through by hand six blades.
23. Propellers High RPM
24. Throttles 1/3 "OPEN"
25. Ignition Switches—"ON" (All Engines)
26. Fuel Pressure—Booster Pump on for Engine to be started
27. Primer (Electric) as required
28. Starter Energizer 30 Seconds, then Mesh—Old Type
28a. Starter Energizer 12 Seconds, then Mesh while still holding Energizer "ON"
29. Mixture—Automatic Lean (after Engine Fires)
    Do not exceed 1400 RPM in "Auto Lean" on Ground

RESTRICTED

BEFORE TAKEOFF:
1. Cabin and Bomb Doors "Closed"
2. Surface Controls—Checked for Freedom & Direction
3. Trim Tabs—Set for "Take-off"
4. Mixtures—Auto Rich
5. Check Fuel Pressure with Booster Pumps "OFF"—"ON" after Check
6. Check Engines at 2000 RPM after head temperature reaches 150° C.
7. Check Switches and Instrument Readings (Vacuum Pressure On—No. 1 and No. 2 Engines—4 ± ¼ at 1000 RPM)
8. Check High RPM Lights "ON"
9. Check Manifold Pressure Full Throttle
10. Supercharger Set and Locked (49" @ 2700 RPM)
11. Wing Flaps—"DOWN" 20° on No. 3 Run Up
12. Cowl Flaps 1/3 "OPEN" (Greater Opening Buffets Tail)
13. Auxiliary Power Unit "OFF"
14. Generators "ON"
15. Landing Gear Lever "DOWN" Position to Check Kick-Out Pressure (should be 825 Lbs. to 875 Lbs.)
16. Gyro Instruments Uncaged
17. Nose Clear of Crew

READ BEFORE TAKE-OFF
AFTER TAKE-OFF AND DURING CLIMB:
1. Gear Up on Pilot's Signal (Brakes Applied to Stop Wheels)
2. Turbo Superchargers Reduced to 45.5"
3. Reduce RPM to 2550
4. Raise Flaps only after gear up lever returns to neutral
5. Fuel Booster Pumps "OFF" (Unless Required for High Altitude Flying)
6. Cowl Flaps adjusted
7. Auxiliary Hydraulic "OFF" on other than local flight

BEFORE LANDING:
1. Notify Crew
2. Nose Clear of Crew
3. Auxiliary Hydraulic Power "ON"
4. Brakes Check Pressure (850 Lbs. to 1050 Lbs.)
5. Automatic Pilot—"OFF"
6. Cowl Flaps—"CLOSED"
7. Mixtures—"Auto Rich"
8. Intercooler shutters "OPEN" (Unless Carburetor is Icing)
9. Booster Pumps—"ON"
10. Wing De-Icer—"OFF"
11. Landing Gear Lever "UP" to Check Kick-out Pressure (Should be 1050 to 1100 P.S.I.)
12. Landing Gear "DOWN" (Air Speed not to Exceed 155 MPH)

Sequence of Operation:
   a. Pressure Builds up Suddenly then Drops
   b. Green Light Turns On

c. Control Handle returns to Neutral
   d. Warning Horn—"OFF" when any Throttle is Closed
   e. Visually Inspect Locks
13. Wing Flaps "DOWN" do not Exceed 155 MPH with Flaps Down
14. Propeller Controls High RPM
15. Turbos (as required) "OFF" Normally (See Supercharger Operation)
16. Landing Gear Latch "DOWN" again (Final Check)

AFTER LANDING:
1. Cowl Flaps "OPEN"
2. Turbos—"OFF"
3. Wing Flaps—"UP"
4. Mixture—"Auto Lean"

TO SECURE AIRPLANE:
1. Engines—"STOPPED"
2. Cowl Flaps "Closed" after Engine Cools
3. Wheels Chocked
4. Surface Controls—Locked
5. All Switches "OFF"
6. Tail Support—Installed
7. Landing Gear Lever in "DOWN" position

OPERATING LIMITS:
Oil Pressure 75-90 PSI
Fuel Pressure 14-16 PSI
Oil Temp. (For Take-Off) 40°-85° C.
Cyl. Hd. Temp. (For Take-Off) 205° C.

Max. Head Temps.
Continuous Operation	232° C.	
Full Power Climb	260° C.	(One Hour Only)
High Speed	260° C.	(One Hour Only)

CLIMB AND HIGH SPEED (Military Power—Available for 5 Min.)
Max. RPM = 2700 and Manifold Pressure = 49" HG up to 23,000 Feet Altitude (Reduce Manifold Pressure 1.5" HG per 1000 Ft. Increase in Altitude over 23,000 Ft.

REQUIREMENTS FOR AUTOMATIC LEAN CRUISING
1. Mixture Controls—Automatic Lean
2. Oil Temperature: 75° C. Max.
3. Oil Pressure: 65 to 100 PSI
4. Cylinder Head Temperature: 232° C. Max.
5. Fuel Pressure: 14 to 16 PSI
6. Cowl Flaps—Closed if possible—or as required
7. Throttle and Supercharger—At Altitudes where Superchargers are required, put Throttles to Stops and Use Supercharger as required. At Lower Altitudes where Supercharger is not Required, bring Throttles back 2" Hg below desired, then use Supercharger to Gain Desired Manifold Pressure.

FLY! DON'T MUSH

With Throttle Cut He Leveled Off And Mushed Along Out of Trim.

The Crew, We Heard, Gave Him the Bird And Wanted to Throttle Him!

POWER PLANT SYSTEM

General—The airplane is equipped with four Pratt & Whitney R-1830-43 radial air-cooled engines having a 16:9 propeller gear ratio. Each engine is equipped with a single-stage, single-speed engine-driven supercharger having a 7.15:1 gear ratio and a General Electric B-2 Turbo Supercharger.

Using 100 Octane fuel, the engine horsepower ratings are as follows:

CONDITION	BHP U.S.	RPM	Approx. Mercury Manifold Pressure	Altitude
Take-off	1200	2700	49.0 in.	
Military	1200	2700	49.0 in.	S.L. to 23,400 ft.
Normal	1100	2550	45.5 in.	S.L. to 25,000 ft.

In addition to the propellers, the following major units are powered by the engines: four P-1 electric generators (one per engine); four fuel pumps (one per engine); one hydraulic pump (No. 3 Engine); and two pumps for vacuum and de-icer systems. (No. 1 and No. 2 Engines.)

Propellers:

Feathering switches above the windshield control feathering motors for each propeller. Two feathering safety devices are used in the circuit breakers to prevent overloading of the feathering solenoid circuits. When the overload conditions are eliminated a push on the circuit breaker closes the circuit. Current limiters in the fast feathering motor power circuit must be replaced when blown.

Cowl Flaps:

Engine cooling is regulated by means of adjustable cowl flaps controlled from the Pilot's Pedestal. Flaps should be operated to maintain correct engine temperatures at all times. The range of operation is 0 to 30° open.

The cowl flaps should be one-third open during take-off and adjustment should be made so as to keep engine temperatures under the specified operating limits.

For cruising and high speed conditions the cowl flaps' setting should be such that the engine temperatures are below the maximum allowable values listed below:

232° C. Continuous Operation.

260° C. Full Power Climb (one hour only).
260° C. High Speed (one hour only).

NOTE: It is recommended that cowl flaps be used from 0 to 1/3 open or 2/3 to full open. Openings of from 1/3 to 2/3 cause tail buffeting under certain conditions.

Open cowl flaps have a definite effect on speed and climb; the added drag and disturbance of the airflow reduce cruising speed approximately eight-tenths of a mile per hour for each degree of flap opening at normal speeds. The opening of cowl flaps has such a serious effect on the climb performance of the airplane that it may not reach much over 23,000 feet with flaps only slightly open. **This is important—do not forget it.**

On landing approaches, cowl flaps are to be closed and are to be opened immediately upon landing.

The engines should never be shut down with the cowl flaps closed regardless of outside air temperature. The residual heat of the engines will burn the ignition wires at the elbows.

Exhaust Gases:

Exhaust Gases are discharged through the waste gate, or operate the Turbo by passing through the blades of the Supercharger turbine wheel.

Carburetors:

Each engine has an injection type carburetor. An electric motor in each nacelle operated from a switch on the pedestal controls a shutter at the Intercooler circulating air duct outlet. Normally, this shutter is open for Turbo Supercharger operation. When warmer carburetor air is needed, due to icing conditions only, the shutter is closed and cooling air is prevented from entering the Intercooler. Icing is indicated by loss of manifold pressure at constant altitude and throttle.

CAUTION: The intercooler shutters are either fully "closed" or fully "open." There is no intermediate position. Hence, the shutters should only be closed during actual icing conditions and remain closed only long enough to eliminate icing, open frequently to check if icing still exists. Prolonged operation with intercooler shutter closed will injure the engine. Watch head temperatures closely while shutters are closed.

Ignition:

Engine ignition is provided by two Bosch magnetos, mounted on the rear face of each engine. The magnetos are controlled by switches on the Co-Pilot's Ignition Switch Panel. Separate switches permit either one or both magnetos to be operated on any engine. Battery switches are on the Co-Pilot's Auxiliary Switch Panel. A master switch bar located just above the magneto switches is available for simultaneously shorting the primaries of all magnetos and breaking battery circuit of the main system.

Starters:

An Eclipse Electric Inertia Starter is installed on each engine. Double throw switches on the Co-Pilot's Electrical Panel energize the starter flywheel rotation and mesh the rotating flywheel with the engine.

Later models are being equipped with a new design of inertia starter which may be kept energized during the crankcase process. However, it must be energized independently for 10 to 12 seconds before meshing.

Old Starter Switch Arrangement: Located on Co-Pilot's Instrument Panel.

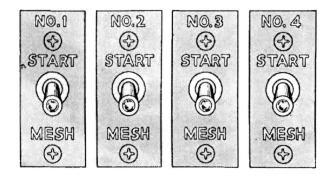

With this arrangement the accelerating motor is not energized on the mesh position.

New Starter Switch Arrangement:

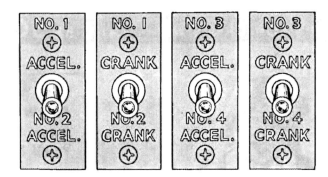

With this arrangement the accelerating motor is a **constant** energizer on the mesh position. Operation is as follows:

Throw "Accelerate" switch on No. 1 Engine and hold until motor is up to speed (10 to 12 seconds). Then, with "Accelerate" switch still **"ON"** throw "Crank" switch on No. 1 Engine. This gives direct cranking action. Repeat on other three engines.

Some airplanes are equipped with nameplates marked "Mesh" instead of "Crank." Either may be encountered.

Power Plant Instruments: All on Co-Pilot's Instrument Panel. (See photo Page 2).

Fuel pressure gauges

Oil pressure gauges

Oil temperature gauges

Engine Cylinder Head Temperature Gauges: Indicate Temperature of cylinder heads.

USE FLAPS FOR CRUISING—AND BE SORRY!

Drag Your Flaps
And Waste Your Gas
Across the Plains
You'll Drag Your Can

On Maneuver
  Dragged His Flaps
Good Thing He Wasn't
  Hunting Japs

Engine Supercharger Pressure gauges—Indicate the pressure in the inlet manifold in inches of mercury. For any throttle setting this pressure may be raised by the degree of turbo supercharger boost.

Tachometers

Lights—Illuminated when the propeller control governors reach upper or lower limits of blade pitch control (not full feather).

Location of Controls: These are in addition to those used in connection with fuel, oil, and supercharger systems:

Throttle Levers	On Pilot's Pedestal
Mixture Levers	On Pilot's Pedestal
Propeller Switches	On Pilot's Pedestal
Propeller Feathering Switches	Over Windshield
Propeller Feathering Circuit Breakers	On Pilot's Pedestal
Cowl Flap Switches	On Pilot's Pedestal
Starter Switches	On Co-Pilot's Switch Panel
Primer Switches	On Co-Pilot's Switch Panel
Battery Switches	On Co-Pilot's Auxiliary Switch Panel
Magneto Switches	On Co-Pilot's Ignition Switch Panel
Master Ignition Switch	Over Magneto Switches
Supercharger Levers' Lock	Left side of Pilot's Pedestal
Throttle Lever's Lock	Right side of Pilot's Pedestal
Push-Pull Meshing Rods	One-in each Nacelle (Not operable in Flight).

Also see "Location of Controls," Page 94.

CONSOLIDATED AIRCRAFT CORPORATION
San Diego, California

April 15, 1942

OPERATING LIMITS FOR R-1830-43 ENGINES (Turbo Supercharger)

Condition	BHP/Eng.	RPM	Man. Pr. Man. Hg.	Alt. Ft.	Mixt.	Max. Cyl. Head Temp. °C.	Max. Oil in Temp. °C.	Required Oil Pres. #/sq. in.
Take-Off (5 Minutes Duration)	1200	2700	49	S. L.	Auto Rich	260	95	80-100
Military (5 Minutes Duration)	1200	2700	49	S. L. to 23400	Auto Rich	260	95	80-100
Normal (Continuous)	1100	2550	45.5	S. L. to 25000	Auto Rich	260 (One hr.) 232 (Continuous)	95	80-100
Maximum Cruising (Continuous)	700	2150	29.5	Any	Auto Lean	232	85 (Auto Rich) 75 (Auto Lean)	65-100

Maximum cylinder barrel temperature   168° C.
Maximum air temperature at carburetor   38° C.
Maximum exhaust temperature at turbo   788° C.
Maximum temperature of ignition system   115° C.
Maximum turbo RPM   21,300
Required fuel pressure   14-17 lb./sq. in.
Maximum dive RPM of engine   3,060 (30 seconds duration)

POWER PLANT – INSTRUCTIONS

NOTE: The following operation instructions are reprinted from the Pratt & Whitney Operating Manual.

Engine—The Pratt & Whitney Twin Wasp R-1830-43 is a 14 cylinder, two row, radial, air-cooled engine, having 16.9 propeller reduction gearing and a single stage, single speed internal engine driven supercharger.

A complete description can be found in the current Operators Handbook for the Twin Wasp C-4 engines.

Carburetor and Mixture Control—The R-1830-43 engine is equipped with the Bendix Stromberg PD-12F2 and PD-12F5 Injection Carburetors. Metering of the fuel is accomplished by air flow through the two carburetor venturis. Four positions of the Pilot's mixture control lever, shown diagrammatically at upper right, are to be used in the operation of the carburetor.

Automatic Rich—The usual operating position of the mixture control, automatically maintains the necessary fuel/air ratio for all flight conditions. The diagram at lower right shows variation of fuel/air ratio with air flow through the carburetor. B.H.P. will very closely correspond to air flow. At high power, the proportion of fuel to air is relatively high, in order to suppress detonation and assist in cooling. Between normal rated and cruising powers, the proportion of fuel is decreased so that in the cruising range, fuel consumption is reduced to the minimum required to prevent detonation and over-heating and to provide good acceleration.

Automatic Lean—Is an alternate operating position of the mixture con-

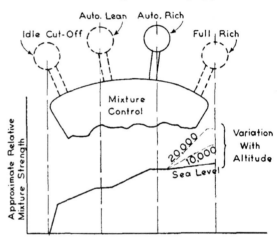

DIAGRAM OF MIXTURE CONTROL QUADRANT
Figure One

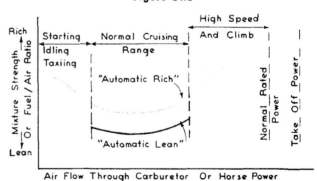

TYPICAL CARBURETOR PERFORMANCE CURVE
Figure Two

trol, resulting in leaner fuel/air ratios than Automatic Rich. The two illustrations on Page 37 show the reduction in mixture strength resulting from changing the control from Automatic Rich to Automatic Lean. During the favorable conditions of stabilized level flight or a cruising descent, Automatic Lean may be used in the cruising power range when fuel economy is of primary importance and when cooling is adequate. However, for conservative operation, Automatic Rich is the preferable mixture control setting.

Full Rich—Full Rich setting of the mixture control renders inactive the altitude compensating device built into the carburetor. Without compensation for density of air flowing through the venturis, the fuel/air ratio will become increasingly richer with altitude. The Full Rich mixture control setting is recommended only when the automatic mixture control unit is believed to be faulty.

Idle Cut-Off—Moving the mixture control past Automatic Lean to the end of its travel will stop all fuel flow, regardless of fuel pressure. Idle Cut-Off is intended principally for stopping the engine without the hazard of backfiring.

As suggested by the graph representing relative mixture strength, plotted beneath the control quadrant in Figure 1 (Page 37), manual adjustment of the mixture control between the four positions will provide a fairly uniform transition in the mixture delivered to the engine. **Manual control, if necessitated by malfunctioning of the automatic altitude compensating unit, should be used cautiously, and leaning of the mixture should not exceed that required for smooth operation.**

Referring again to Figure 2 (Page 37), mixture strength is increased below the cruising power range. This enrichment provides easier starting and the dependable acceleration needed in taxiing and the approach for a landing. Fuel metering in this power range is accomplished largely by throttle opening.

The accelerating pump is operated by, and in proportion to, the momentary changes in air pressure in the supercharger entrance. The accelerating pump is not connected with the throttle or throttle controls. Hence, when the engine is not running, no fuel is pumped from the carburetor when the throttle is moved, no matter how rapidly.

Mixture Control Quadrant Requirements—The mixture control quadrant should have a stop or movable catch at the "Automatic Rich" position." "Full Rich" is seldom used and is provided only to care for emergency operation in case the automatic unit is not functioning. A second stop or catch is desirable at the "Automatic Lean" position to guard against inadvertent movement toward the "Idle Cut-Off" position.

Starting—With Electric Inertia Starters—Where engines are warm from previous running or where outside air temperatures are 60° F. (15° C.) or above, only 4-5 seconds priming will be necessary.

1. Have the throttle approximately 1/5 open (after the springiness of the control system has been taken up).

2. Mixture control in the "Idle Cut-Off" position.

3. Turn Ignition Switches to "Both ON."

4. Be sure booster pump is "OFF" after priming.

5. Close "START" or "ACCEL" switch, then throw to "MESH" or "CRANK" which will simultaneously energize the booster coil.

6. Leave mixture control in "Idle Cut-Off" until engine fires. The reason for this precaution is that the high speeds obtained with these starters can cause the engine fuel pumps to develop pressures in excess of 5 lbs./sq. in., irrespective of whether the wobble pump is operated or not. Should the mixture control be placed prematurely out of "Idle Cut-Off" and the engine fail to start for one reason or another considerable quantities of fuel may be discharged into the engine and drain from the blower section, creating a fire hazard.

7. If a start is not effected in a reasonable length of time, investigation should be made to ascertain the cause. As it is impractical to use the mixture control to obtain partial priming as is done with direct cranking starters, it is advisable to use only the hand primer under any conditions where it is felt that the engine is not getting sufficient fuel to start. Overloading will be indicated by a discharge of fuel from the drain located in the lower part of the engine blower. In this case, **keep the mixture control in "Idle Cut-Off,"** open the throttle and turn the engine over with the starter in order to clear it out. If the engine has been loaded and the ignition is left on, it is frequently possible to effect a start while clearing out the engine with the starter. In this case, it is necessary to be ready to immediately retard the throttle to prevent over-speeding. If the ignition switch is not left on during the clearing-out procedure, a reasonable number of turns, such as 6 or 8 revolutions of the propeller, should be sufficient to clear. Then repeat the starting procedure outlined above.

8. If no drainage of fuel is evident from the engine blower, the difficulty is probably not from overloading. In this case it is possible that the engine has not yet obtained sufficient fuel due either to insufficient priming or to the mixture control not having been moved out of "Idle Cut-Off" soon enough. In this case, repeat the starting procedure, operating the primer and the mixture control with caution so as to feed a little more fuel into the engine.

9. If it is still not possible to start the engine, in all probability some part of the ignition system is not functioning, such as the booster coil, and investigation should be made. Protracted operation of the booster can sometimes overheat the coils so as to render the booster inoperative.

10. As soon as the engine fires and before the prime is used up, move the mixture control without hesitation to the "Automatic Rich" position.

11. Adjust throttle control to hold the engine to as low a speed as possible for the first 30 seconds after starting and watch for an indication of oil pressure on the gauge.

CAUTION: If oil pressure does not register on the gauge almost immediately, STOP and investigate.

12. After the first half minute, adjust the throttle to about 1000 RPM propeller in High RPM (Low Pitch).

Priming—When the engines are cold and have been exposed to outside air temperature below 60° F. (15° C.), considerable priming and larger throttle openings will be necessary.

The lower the temperature, the greater the amount of priming which will be required. Under the various temperature conditions which may be encountered, experience will indicate how much priming and throttle opening is necessary to obtain good starting.

Excessive priming will load the cylinders with raw gasoline making it difficult to start the engine. Excessive priming also has a tendency to wash the oil off the cylinder walls and may result in barrel scoring or piston seizure. Rusting of piston rings and cylinder walls will occur if the engine is allowed to stand for a day or more after unsuccessful attempts to start, unless the surfaces are protected by a fresh application of oil. Under priming is usually indicated by backfiring of the engine through the intake system with attendant hazards.

> NOTE: On cold engines, overloading is not necessarily indicated by a discharge of fuel from the engine blower but rather by the presence of liquid gasoline in the exhaust collector and piping. In this case, follow the procedure outlined for clearing out a warm engine when loaded.

If there is no odor or vapor of gasoline in the exhaust, in all probability the engine has not been given sufficient prime, **even though fuel may be draining from the blower.** In cold weather considerable quantities of fuel may be discharged into the blower and pass out through the drain and still leave the engine underprimed. The reason for this is that fuel at low temperature discharged into the blower is not sufficiently vaporized to be carried into the cylinders in mixture strengths necessary for combustion when the engine is turned over. For this reason, direct priming to the intake port is required in cold weather. On the other hand, in warm weather, both the fuel and the engine are at higher temperatures so that fuel discharged into the blower is vaporized sufficiently to be carried into the cylinders in mixture strengths necessary for combustion when the engine is turned over. When underpriming is suspected, additional priming should be made cautiously and the starting procedure repeated as outlined for warm engines.

Warm-Up—After the first half minute, the warm-up should be made with the propeller, regardless of type, in the "Low Pitch" or "High RPM" position and at an engine speed of about 1000 RPM.

Be sure the cowl flaps are open. **Do not attempt to warm the engine up more quickly by closing the cowl flaps at any time,** even in extremely cold weather. This may cause burning of the ignition system at the spark plug elbows.

The oil pressure relief valve has been fitted with a temperature control that forces the oil when cold through the engine under high pressure, as much as 300 lb./sq. in. or more when very cold. This extra high pressure is reduced when an oil-in temperature of about 40° C. (104° F.) is obtained.

Avoid operating above 1000 RPM until the oil temperature has exceeded 40° C (104° F.) and cylinder head temperature reached at least 248° F.

Long continued idling below 800 RPM may result in fouled spark plugs; however, with an occasional run-up for cleaning out, short periods of idling at 400 or 500 RPM are readily usable.

Ground Test—When the oil-in temperature has risen above 40° C. (104° F.), the throttle may be opened to approximately 30" Hg absolute manifold pressure with the propeller in

"Low Pitch" or "High RPM." Note the loss of revolutions when switching to one magneto at a time. In switching from both magnetos to one, the normal drop-off is 50 to 75 RPM and does not usually exceed 100 RPM. When switching from one magneto to the other, the change in RPM should not be more than 30 or 40. It should be noted that the loss in RPM when operating with one or two magnetos varies with different engine speeds. This check should be made in as short a time as practicable. Continued running on one magneto with manifold pressure as high as 25" to 30" Hg absolute may cause serious detonation.

Noticeable vibration of the engine relative to the airplane structure will usually result when one or more cylinders are misfiring due to a malfunctioning spark plug. At low engine speeds on either or both magnetos, freedom from vibration is an excellent indication of proper functioning of the engine, particularly of the ignition system.

In rare circumstances, even after the engine has been run a sufficient length of time to give reasonable assurance that the spark plugs are cleared out, excessive RPM drop or uneven engine operation may be experienced during the regular magneto check.

In this case it is permissible to make a quick check of magnetos at 33" Hg in low pitch in order to determine if the trouble lies in the magnetos themselves.

> CAUTION: Operation on one magneto at this power output must be held to the shortest possible length of time because of the possibility of serious damage from detonation.

Check oil pressure, oil temperature, fuel pressure, and other items at 2000 RPM. Note the manifold pressure as a reference for checks in the future.

Oil pressure measured at the pressure gauge take-off on the right hand side of the engine rear section should be 85+15−5 lbs./sq. in. at 2000 RPM with 60° C. (140° F.) oil inlet temperature. Oil pressures will vary with RPM and temperature and need cause no alarm by falling to as low as 15 lbs./sq. in. with the engine idling or if the pressure rises somewhat over 100 lbs./sq. in. with cool oil at take-off RPM.

Fuel pressure should be 15 ± 1 lb./sq. in. relative to carburetor air pressure. The fuel pressure at low idling speed may be as low as 8 or 10 lbs./sq. in. and still be satisfactory if the pressure comes up to the desired amount with 800 or 1000 RPM.

The suctions and back pressures of the oil system in the airplane often are different from those of the test stand upon which the engine was run. Consequently, engine oil pressure, obtained initially in the airplane installation, may require adjustment of the main oil pressure relief valve. On subsequent running of the engine, any appreciable change in oil pressure under the same condition of RPM and oil temperatures may indicate trouble within the engine or oil system which should be investigated.

Cooling of the cylinder heads and barrels, and ignition harness is usually insufficient while on the ground for continued running above 1400 RPM to 1500 RPM. Avoid prolonged running at power above this. It is recommended not to exceed 232° C. (450° F.) head temperature during ground operation.

Cowl Flaps—The adjustable cowl flaps or gills shall be fully opened during all ground operation and at least partially opened for take-off and climb. During flight the cowl flaps should be adjusted to keep the cylinder temperatures under the limits specified.

Taxiing—Taxiing is normally accomplished with the engine and accessory controls in the positions usual for take-off except the throttle should be practically closed, and the cowl flaps wide open. The throttle is used to regulate power for movement and control of the airplane. A steady flow of power is preferable to repeated short bursts which tend to empty the accelerator pump if sufficient time is not allowed for the pump to refill.

Flight—General—In general, the ratings of the engine have been established as near the high limits as possible, consistent with the fundamental concept of dependable power during a reasonable period between overhauls. These engine ratings are limited or bounded by three main criteria: Brake horsepower, brake mean effective pressure, and speed, along with others concerned with operating conditions, such as temperatures. These ratings as set forth in the engine specification may be defined as follows:

Take-off Rating—This is the maximum power and engine speed permissible for take-off and should be maintained only long enough to clear obstructions.

Military (5 Min.) Rating—This is the maximum power permitted for the military services with less regard for long life of the engine than for immediate tactical needs. Military ratings comparable to "Take-Off" power with manifold pressures modified to suit altitude conditions may be used for five minutes duration in any attitude of flight.

This is solely a military consideration and is not approved for commercial operation.

Normal Rated Power—This is frequently referred to as a Normal Maximum Rating or Maximum Except Take-Off power ("METO POWER") and is the maximum power at which an engine may be operated continuously for emergency or high performance operation in climb or level flight. It can be maintained up to 25,000 feet. Above that altitude the manifold pressure should be reduced 1.5 in. per 1000 feet increase in altitude to avoid overspeeding the exhaust turbine.

Maximum Power and RPM for Cruising—This rating stipulates both the maximum power and maximum RPM permissible for continuous operation with the mixture control in Automatic Lean.

With constant speed propellers an infinite number of engine RPM and manifold pressure combinations are available for any desired value of engine power or airplane speed. In some airplanes, undesirable vibration may be induced in the airplane structure or propeller at certain engine speeds, and these speeds should be avoided or passed through quickly. The proper combination of RPM and manifold pressure for the particular horsepower desired can be determined from the following pages of this instruction or from the Air Corps Cruising Control Chart.

Except to prevent carburetor icing and congealing of oil in the turbine regulator lines, **the turbine supercharger will not be used until the engine throttle is opened to the ¾ position or against the stop,** if one is provided. After the throttle is opened to this position, engine power will be regulated by the propeller control and turbine regulator.

As the airplane climbs, the engine intake manifold pressure may be kept constant by the gradual opening of the throttle or adjustment of the regulator control. In no event should the engine be supercharged more than the specified pressure recommended for the engine as pre-ignition may occur, and the engine may be seriously damaged if this condition is continued.

In general, to reduce the power of the engine, first reduce the supercharger regulator setting in order to slow down the turbine and to reduce the carburetor pressure. Do not throttle the engine completely until the regulator control is in the lowest position. To increase power, the converse is used: first open the throttle and adjust the propeller governor, then gradually, or as fast as the power plant will respond readily, advance the regulator control to obtain the desired manifold pressure, being sure that the turbine control has not been advanced so fast or so far that the pressures will become excessive when the turbine supercharger has arrived at its regulated operating pressure.

In the event of erratic or abnormal operation of the engine or engine instruments, the regulator control should be immediately placed in the lowest pressure position.

In descending from high altitudes, the supercharger control should remain at an automatic position; i.e., not at the lowest position.

Under normal atmospheric conditions the intercooler shutter controls are left in the "cold" position, being used only when icing conditions occur in the carburetor.

General smoothness, engine speed, manifold pressure, carburetor air temperature, fuel/air ratio, cylinder temperatures, oil temperature, and the oil pressure give the most satisfactory indication of the performance of the power plant. If any of these appears irregular, the engine should be throttled, and, if the cause is not apparent, a landing should be made to investigate and correct the trouble.

Climb—Climbs are normally made with less than normal rated power, usually about 75%. The tactical requirements of the military services may dictate using military rated power in climb for a period not exceeding five minutes or normal rated power for longer periods.

Climbs should be made with "Automatic Rich" mixture control.

Adjust the cowl flaps to maintain cylinder temperatures somewhat less than the maximum permissible, preferably about the maximum permissible for cruising, i.e., 232° C. (450° F.).

A material reduction in cylinder and oil temperatures can be obtained by climbing at an indicated air speed ten or twenty miles per hour higher than the speed for best climb without much loss in rate of climb.

A tendency for the oil to overheat can be checked the quickest by reducing the engine speed rather than by throttling alone.

Some small adjustment of the turbine control may be necessary during the climb to regulate the manifold pressure to the desired value. It will be noted that for a constant regulator setting, and a given RPM and engine throttle setting, the manifold pressure will increase slightly with increase of altitude.

Cruising—Maximum—The Cruising—Maximum entry in the "Pilot's Check Chart" is the highest power and RPM which may be maintained continuously with Auto Lean mixtures. Cruising—Maximum utilizes the highest permissible cruising RPM in order to maintain the maximum permissible cruising power to the highest possible altitude and should be reserved for adverse conditions. Average cruising requirements normally call for reductions of 15% to 30% from maximum cruising power and still further reductions where long range or endurance is a requisite and where maximum fuel economy is desired. Engine maintenance economy and long periods between engine overhaul are largely a function of using conservative cruising powers.

After the airplane has leveled off, and while attaining its approximate cruising speed, the engine should be given an opportunity to cool down, preferably even below the final cruising temperatures, before the carburetor is changed to the "Automatic Lean" position. This permits the blower and rear sections, as well as the cylinders, to cool down. A well cooled engine is less likely to detonate when the mixture is leaned than a hot engine.

To aid in the cooling of the engine as outlined above, the cowl flaps should not be closed to the minimum position immediately after completion of the climb, but progressively as the airplane gathers speed. It is desirable to maintain cylinder head temperatures below 204° C. (400° F.) in level flight while cruising, and under no circumstances should they be allowed to exceed 232° C. (450° F.).

Cruising—General—For obtaining the maximum economy from the airplane as a complete unit it is recommended that cruising power be controlled in accordance with the "Cruising Control Chart Model B-24D" published by the Air Corps. This chart enables the operator to determine the most efficient power setting (manifold pressure and RPM) for the operating conditions (outside air temperature, altitude, gross weight, and desired airspeed). It also gives the fuel flow for this setting.

At high altitudes (20,000 ft. and above) it will not be practicable to operate at engine speeds below 1800 RPM. At lower speeds there is not a sufficient quantity of exhaust gas to operate the turbo and consequently cruising manifold pressures cannot be maintained.

For low-altitude, low-power cruising operation, it may often be possible to obtain the desired manifold pressure without use of the turbo supercharger. In order to keep the turbo engaged the following procedure is recommended:

1. Engage turbo supercharger and increase the manifold pressure approximately one inch.
2. Reduce engine throttle opening to maintain the original manifold pressure.

Cruising Descent—Under normal conditions of cross country flight it is general practice to start a descent at a distance from the destination of as much as 100 miles or more. This distance is determined as a function of the altitude of the airplane above the destination, the rate of descent desired and the time necessary for the descent, the wind velocity and direction and its effect upon the airplane speed, and the resulting speed of the airplane during the descent. Such a descent should be regarded as a cruising operation. Cruising RPM, power, and mixture should be maintained throughout the descent until the point is reached where the final glide or approach for landing is to be made.

Dive—The centrifugal or inertia loads on the master rod bearing increase as the square of the RPM. These loads, however, are in the opposite direction from the power impulse loads from the pistons. Therefore, high engine speeds with low manifold pressures impose the severest loads on the master rod bearings and should be avoided if possible. Where overspeeding of the engine is unavoidable in dives, it is recommended that the throttle be partially opened to give 12" to 15" Hg. if practicable. However, this may increase the diving speed of the engine somewhat. The maximum safe overspeed RPM has been defined at 3060 RPM.

Since dives are usually accompanied by other maneuvers that may require full power of the engine, the mixture control must be in the "Automatic Rich" position.

The turbine regulator control may be left in the position it was in when entering the dive, or it may be reset according to the power desired after the dive.

Glide and Approach for Landing—The paragraph on **Dives** has bearing upon the transition from high speed flight or cruising to the throttled condition of a glide or approach for landing. While the airplane speed is being reduced, the propeller governor should be set for maximum cruising RPM or less to prevent high speed "windmilling" of the engine. The throttle should be closed as desired.

When the airspeed has been satisfactorily reduced and the airplane is in position for the final approach, make the following adjustments or checks:

1. Propeller control in automatic and to a setting between maximum cruising and normal rated engine speed, preferably near cruising. This position permits a more rapid opening of the throttle without danger of serious overspeeding than does the take-off setting of the propeller and gives up to, or more than, normal rated power for emergency. Full take-off power is available by adjustment of the propeller control after the throttle has been opened.
2. Fuel selector valve on tank with sufficient fuel.
3. Mixture control in "Automatic Rich."
4. Turbine regulator set approximately at the take-off position.

Unless the landing characteristics are adversely affected, it is advisable to partially open the cowl flaps, particularly if slow level flight is maintained after the landing gear or wing flaps are lowered. If emergency power is needed, further adjustment of the partially opened flaps can be made after having cared for more urgent duties.

Before taxiing, have the cowl flaps fully "OPEN," and turbine regulator "OFF."

Stopping—If the cylinders are hot due to hard taxiing, permit the engine to idle a short time to allow the cylinder temperatures to cool below 205° C. (400° F.).

If a cold weather start is anticipated, the oil dilution system should be used.

Leave Hydromatic and electric propellers with the controls set in the take-off or High RPM position.

To stop the engine, move the mixture control to the "Idle Cut-Off" position. This may be done with the engine turning at any idling speed. When the engine has stopped, turn all ignition switches to "OFF."

If "Idle Cut-Off" should not stop the engine, close throttle, cut switch and **slowly** open throttle wide. Have "Idle Cut-Off" adjusted properly as soon as possible.

After the engine stops, leave the cowl flaps fully open to aid in circulation of air over the engine. This precaution is to guard against residual heat of the power section and exhaust

PILOT'S CHECK CHART

Pratt & Whitney Twin Wasp R-1830-43; Single Stage, Single Speed: PD-12F2 and PD-12F5; Fuel: Aviation Grade 100

Operating Condition	Max. RPM or Gov. Setting	Max. Man Pressure in Hg.	Mixture Control	Oil Press. Limits Lbs./In.^{2}	Oil in Temp. Limits °C.	Cyl. Heads Max. °C.	Cowl Flaps	Approx. Fuel Cons. U.S. Gal./Hr.
(Pull Through) Start	700	1/5 Throttle	Idle Cut-off then Auto Rich	Show in 30 sec. (400)				
Warm-Up	1000		Auto Rich				Full Open	
Ground Test	(*High RPM) 2000	30.0	Auto Rich	80-100	40-70	205	Full Open	
Take-Off	2700	48.0	Auto Rich	80-100	40-85	260	Part Open	150
Military (5 min.)	2700	48.0**	Auto Rich	80-100	40-100	260	Part Open	150
Rated Power	2550	43.5**	Auto Rich	80-100	40-85	260	As Req'd	132
75% Rated	2230	34.0	Auto Rich	80-100	40-85	232	As Req'd	72
Cruising—Max.	2230	31.0	Auto Lean	65-100	60-75	232	Closed	56
Dive	3060	(12-15)	Auto Rich		60-75	232	Closed	
Glide and Descent	(*Cruise RPM)		Auto Rich		60-75	232	Closed	
Approach for Landing	(*2300)		Auto Rich		60-75	232	Part Open	
Stop	(*High RPM)		Idle Cut-Off	15 (Min. Idling)		205	Full Open	

* Propeller Governor Setting.
** **Note:** Reduce manifold pressure 1.5 in. per 1000 ft. above 25,000 ft. to prevent overspeeding of turbine supercharger.
Desired Fuel Pressure: 15 ± 1 lbs./sq. in.   Desired Oil Temp.: 60-75° C.

collector damaging ignition system insulation by raising temperatures above the safe limits for the spark plug elbows (120° C. or 248° F.).

Leave the mixture control in "Idle Cut-Off" position at all times when the engine is not running. If the fuel supply is located higher than the carburetor, turn the fuel selector valve "OFF" particularly if the engine is not to be used for two or three days or longer.

Specific Operating Instructions—Twin Wasp R-1830-43 Single Stage, Single Speed Supercharger, 7.15:1 Blower Ratio; PD-12F2 or PD12-F5 Stromberg Injection Carburetor; Bosch SF-14LU-6 or -7 Magnetos; .5625 (16.9) Propeller Reduction Gearing; Fuel: Aviation Grade 100.

Ratings	BHP	RPM	Altitude
Take-off	1200	2700	
Military (5 min.)	1200	2700	S.L. to 25,000'
Normal	1040	2550	S.L.
	1100	2550	6200 to 25,000'
Cruise—Maximum	735	2230	
Maximum Diving Speed		3060 (30 seconds)	

Cylinder Temperatures—Maximum	Heads*		Barrels**	
	°F.	°C.	°F.	°C.
At rated take-off and military power (5 min.)	500	260	335	168
High speed and climb at normal rated power	500	260	335	168
Continuous operation at any power except as above	450	232	335	168
Desired continuous operation (LESS THAN)	400	200	275	135

*Measured at point embedded in gasket of rear spark plug.
**Measured on rear side at point embedded in fillet of hold down flange.

Oil Pressure	lbs./sq. in.	
Desired, at 2000 RPM at 140° F. (60° C.)	85+15, −5	
Minimum at 2550 RPM at 100° C.	80	
Minimum at 2230 at 85° C.	65	
Minimum at 1200 at 85° C.	45	
Minimum at Idling	15	

Oil Inlet Temperatures	°F.	°C.
Minimum for take-off and flight	104	40
Desired	140-167	60-75
Maximum, level flight	185	85
Maximum, climb or military	212	100

Fuel Pressure	lbs./sq. in.
Desired	15 ± 1
Allowable range	14 − 16

These fuel pressures are based on the assumption that the fuel pump and pressure gauge are vented to the carburetor air scoop.

TURBO SUPERCHARGER

General—The power which an internal combustion engine develops decreases as the pressure of the charge entering the cylinders decreases. The density (weight per unit volume) of the atmosphere decreases with increase in altitude. The pressure in the cylinders of an unsupercharged aircraft engine will, therefore, decrease with altitude, with a corresponding reduction of engine power. The function of the supercharger is to overcome this loss of power by supplying air to the engine, at or above sea level pressure (27" to 30" Hg.) from sea level to critical altitude of 25,000 feet.

The turbine supercharger is a centrifugal compressor which derives its operating power from the exhaust of the engine. The engine exhaust gas is conducted to a nozzle box and directed against buckets on the turbine wheel.

The impeller or rotor, on the same shaft as the turbine wheel, is the only major moving part. The speed of the rotor is controlled by a waste-gate in the exhaust system which controls the pressure of the exhaust against the impellers. The speed control is, therefore, very accurately controlled.

The engine supercharger would normally be sufficient to give full take-off power at sea level, but because of the intercoolers and ducts it is necessary to use the turbo superchargers for take-off to overcome the air drag within the installation. In climbing from sea level to altitude more and more boost is required (due to decreased atmospheric pressures 30" Hg. at sea level to 12" Hg. at 25,000 feet) in order to maintain approximately sea level pressure on the carburetor inlet.

If a flight is to be made above critical altitude it must be made at a sacrifice in horsepower. (The manifold pressure reduction will be 1½" Hg. per 1000 feet above 25,000 feet.) This reduction of manifold pressure will keep within the maximum turbine RPM limits.

Because of exhaust back pressure, the engine with the exhaust-driven turbo supercharger must run at a slightly higher manifold pressure in order to obtain the same indicated or brake plus supercharger horsepower requirements.

If the engine RPM is reduced excessively the turbine will have insufficient gases to work with and a complete collapse of the cycle may occur. This type of operation gives the impression of improper supercharger regulation and in cases where this occurs the engine RPM should be increased.

Intercoolers—Heat of compression of air by the superchargers must be taken off before it reaches the engine; otherwise, the normal carburetor intake temperature limits will be exceeded. This is accomplished by intercoolers or radiators in the air intake duct between the turbo supercharger and the carburetor. Shutters on the intercooler are provided to regulate the carburetor air temperature to prevent icing. Intercooler shutters should only

**Pilot Pitt Flew At Great Height
With Intercooler "Closed"
Which Was Quite Right.
On Coming Down, Forgot...
(The Fool)
That Intercooler "Open"
Helps Her Cool.**

HOT AIR = HOT MOTORS

be used when it is evident that the engine is losing power due to icing in the induction system. Intercooler shutters are two position—"FULL OPEN" and "FULL CLOSED." Extreme caution should be exercised when using intercooler shutters, and cylinder head temperatures must be watched closely.

Regulators—Superchargers are controlled through a supercharger regulator which automatically controls the waste-gate to maintain a constant exhaust pressure setting on the nozzle box. The regulator, once set for any given pressure within a range of from 28" Hg. to 52" Hg. will keep the manifold pressure constant unless the altitude or speed are changed.

Carburetors—Due to the construction of the Bendix fuel injection carburetor, the superchargers can change the fuel air mixture ratio slightly. Therefore, better engine operation and fuel consumption will result if the turbo superchargers are set so that they boost at least two inches of the desired manifold pressure.

Controls—The supercharger controls are the left unit of the three groups of carburetor controls on the pedestal. Four control levers operate the hydraulic waste-gate operating valves on the four engines.

The speed of the turbo supercharger is reflected in the manifold pressure. The manifold pressure gauge should be consulted and the supercharger regulator control manipulated in the same manner as the throttle.

NOTE: The turbo supercharger should be kept engaged at all times (except in final approach to land) not only as a means of preventing induction system icing but also to insure a continuous supply of warm oil to the turbo supercharger and to the regulator. Even when the desired operating manifold pressures may be obtained by the use of the engine supercharger alone, it is desirable to engage the turbo sufficiently to provide approximately 2 inches manifold pressure.

Operating Instructions—See "Airplane and Engine Operating Procedure."

WARNING: Turbo controls should be in the "OFF" position in the final approach for a landing, except that in emergency when full power is anticipated or landing is being made at altitude, for this reason: When the throttles are closed the waste-gate would also close due to lack of exhaust pressure and any backfiring would result in blowing off the exhaust installation. In any case, put in "OFF" position immediately after landing and be careful to advance throttle slowly if opened while turbo control is "ON."

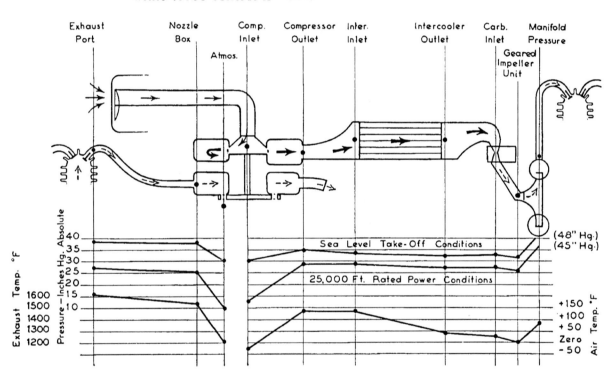

SUPERCHARGER PRESSURE AND TEMPERATURE CYCLE

The performance or cycle throughout the supercharger installation is diagrammed above. The pressure cycle is shown in the upper two curves, while the temperature cycle at 25,000 feet is shown in the lower curve.

The black dots indicate the points in the installation to which the ordinates are referenced. The cycle follows the exhaust gas from the cylinder exhaust port through the turbine to atmosphere, then the air cycle is picked up and proceeds from the compressor inlet through the compressor, intercooler, carburetor, and geared supercharger to the cylinder intake manifold.

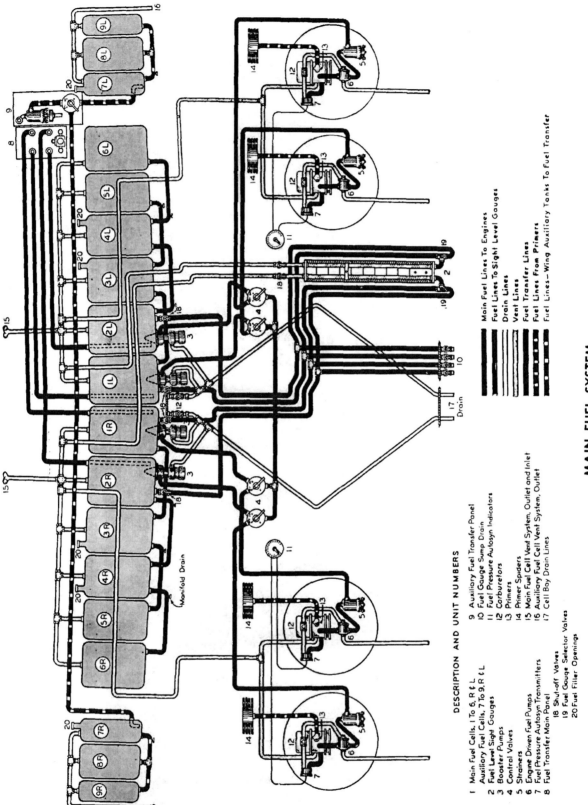# MAIN FUEL SYSTEM
B-24D

DESCRIPTION AND UNIT NUMBERS

1. Main Fuel Cells, 1 To 6, R & L
2. Auxiliary Fuel Cells, 7 To 9, R & L
3. Fuel Level Sight Gauges
4. Booster Pumps
5. Control Valves
6. Strainers
7. Engine Driven Fuel Pumps
8. Fuel Pressure Autosyn Transmitters
9. Fuel Transfer Main Panel
10. Auxiliary Fuel Transfer Panel
11. Fuel Gauge Sump Drain
12. Fuel Pressure Autosyn Indicators
13. Carburetors
14. Primers
15. Primer Spiders
16. Main Fuel Cell Vent System, Outlet and Inlet
17. Auxiliary Fuel Cell Vent System, Outlet
18. Shut-off Valves
19. Fuel Gauge Selector Valves
20. Fuel Filler Openings

Main Fuel Lines To Engines
Fuel Lines To Sight Level Gauges
Drain Lines
Vent Lines
Fuel Transfer Lines
Fuel Lines From Primers
Fuel Lines—Wing Auxiliary Tanks To Fuel Transfer

RESTRICTED

FUEL SYSTEMS [Page 51

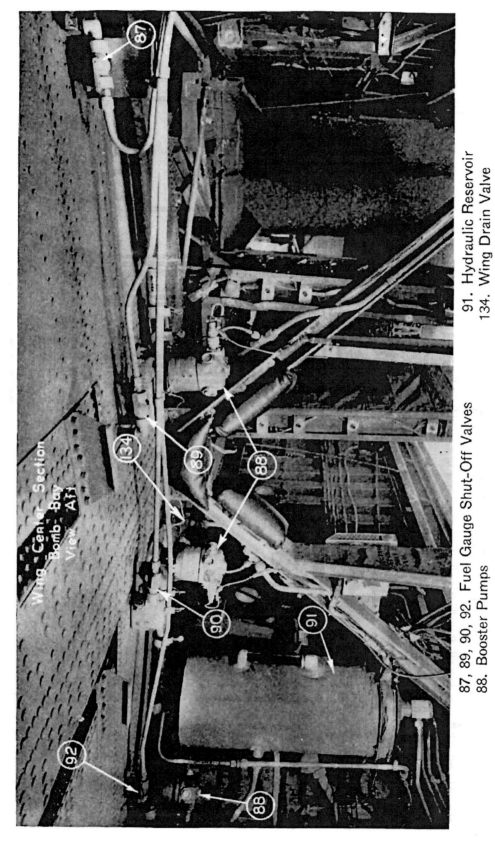

91. Hydraulic Reservoir
134. Wing Drain Valve
87, 89, 90, 92. Fuel Gauge Shut-Off Valves
88. Booster Pumps

FUEL SYSTEMS

General: The following fuel capacity is provided:

Main System, 12 cells, 4 systems .... 2343 U. S. Gal.

System 1	617 U. S. gallons
2	557 U. S. gallons
3	556 U. S. gallons
4	613 U. S. gallons

Auxiliary Wing Tank System, 6 cells, 2 systems .... 450 U. S. Gal.

System 1	225 U. S. gallons
2	225 U. S. gallons

In flight, fuel from auxiliary systems should be transferred to main system as soon as possible to improve loading conditions and reduce fire hazards.

Main Fuel System: Comprises:

1. Twelve self-sealing fuel cells in the wing center section. There are four sets of three cells each. In normal operation each engine is served by one set.

2. Four electrically-driven booster pumps with strainers (one for each set of cells). They are in the bomb bay just under the cells.

3. Four triple-port shut-off valves.

   On each valve: One port leads to an engine; one port leads to a set of cells; and one port interconnects to the other three valves. The "cross-over" connection

95. Fuel Selector Shut-Off and Crossover Valve

allows fuel from any set of cells to serve any engine in an emergency, and permits equalizing flow between systems. These valves are under the front spar in the bomb bay.

4. Four engine-driven pumps with strainers are located—one in each nacelle.

5. Four electrically controlled primers are located—one on each carburetor.

6. Two venting systems. One serves the six cells (two sets of three each), one fuel gauge, and the two carburetor vapor chambers on the right side of airplane centerline; the other serves similarly on the left. Two outside openings, one of scoop, near the junction of the fuselage and wing, and the other an outlet aft of the scoop serve each vent system. The scoop opening serves two purposes: it rams air into cells to keep pressure slightly above atmospheric; its relief valve allows pressure to escape when airplane is at rest and when the cell pressure reaches 0.3 p.s.i. (0.2 Kg./sq.cm.). The second valve is solely a relief valve, with a flush outlet.

SPECIAL NOTE: a. Should both valves fail, atmospheric venting (but not ramming) may be obtained by disconnecting the vent lines at the vent valves and allowing cells to vent inside fuselage.
b. On future airplanes and, as soon as possible, on present airplanes, newly designed outlets will be installed which will replace the Kenyon valves.

93. Fuel Transfer Pump Panel
94. Life Raft Cradles

7. One electrically driven transfer pump allows transfer of fuel from one set of cells through the transfer panel.

8. Wing drain and booster pump vent lines empty overboard under bulkhead at Station

4.0. Provides drainage for booster pumps and for wing cell compartments. The two shut-off valves are normally open but must be closed during combat.

9. Main fuel system supply lines and cell interconnecting, or manifold lines are self-sealing.

Auxiliary Wing Tank Fuel System: Comprises:

1. Six self-sealing fuel cells. Three in each wing, outboard of the wheel well.
2. One 2-way selector shut-off valve which may be connected to the main transfer pump. This allows auxiliary wing tank fuel from either set of 3 cells to be transferred into any one of the four sets of cells in the main system.
3. Two venting systems, one for each set of three cells, right and left. The characteristics are similar to the main venting systems, except that each vent has only one opening.
4. Auxiliary wing cell fuel supply lines are self-sealing.

Fuel System Indicators:

Pressure—Gauges measuring fuel pressure at the carburetors are mounted on the Co-Pilot's Instrument Panel. (See photo Page 2.)

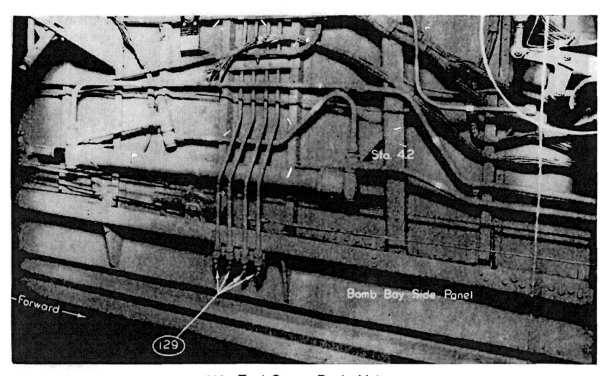

129. Fuel Gauge Drain Valves

Quantity—Sight gauges, mounted on the forward face of bulkhead at Station 4.0 show the quantity of fuel in each unit of the system. In case of damage to the gauge vent or supply lines, shut-off valves are provided on top of the gauges and at the supply take-off under the center section, to prevent the loss of fluid. (See Pages 51 and 52.)

No fuel quantity gauges are provided in either auxiliary system. A glass tube between the wing cell selector valve and the transfer pump shows flow of gasoline being transferred.

NOTE: Inclinometer on inboard side of water jug must read zero when gauge readings are taken.

81. Ammeters
82. Voltmeter Selector Switch
83. Generator Switches
84. Vacuum Selector Valve
85. Voltmeter
86. Fuel Gauges

Page 56] FUEL SYSTEMS   RESTRICTED

Location of Controls.

Main Booster Pump Controls	On Co-Pilot's Switch Panel
Main System Shut-off and Cross-Over Valves	Under front spar in bomb bay. No. 1 and No. 2 at left, No. 3 and No. 4 at right.
Transfer Pump	Aft of Radio Compartment over bomb bay.
Engine Primer Controls	On Co-Pilot's Switch Panel
Main System Vent-disconnect Lines	To right and left of fuselage forward of transfer pump.
Auxiliary Wing System Shut-off and Selector Valve	Aft of transfer pump.
Fuel Gauge Line Shut-off Valves	Under main fuel tanks in bomb bay, forward of booster pumps.
Fuel Gauge Vent Shut-off Valves	Over fuel gauge.
Wing Drain Line Shut-off Valves	Under main fuel tanks in bomb bay, aft of booster pumps.
Bomb bay Tank Selector Valve and Booster Pump	Between bomb bay tanks on the catwalk.

See also "Location of Controls," Page 94.

Bomb Bay Tanks

Bomb Bay Tanks are provided for auxiliary installation. These tanks are equipped with a single two-way valve which will take suction from either or both tanks. The outlet leads to a C-10 pump which will transfer the fuel either to a "T" on the main valve cross-connection or to the fuel transfer panel for distribution to the main tanks. Both installations are used.

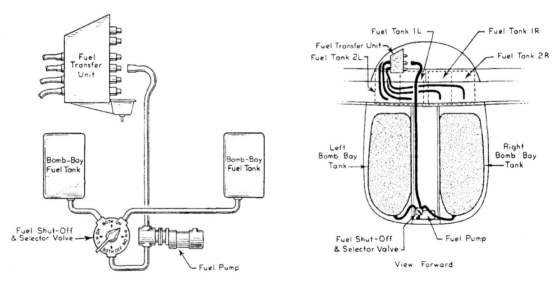

AUXILIARY FUEL SYSTEM—BOMB BAY
B-24D

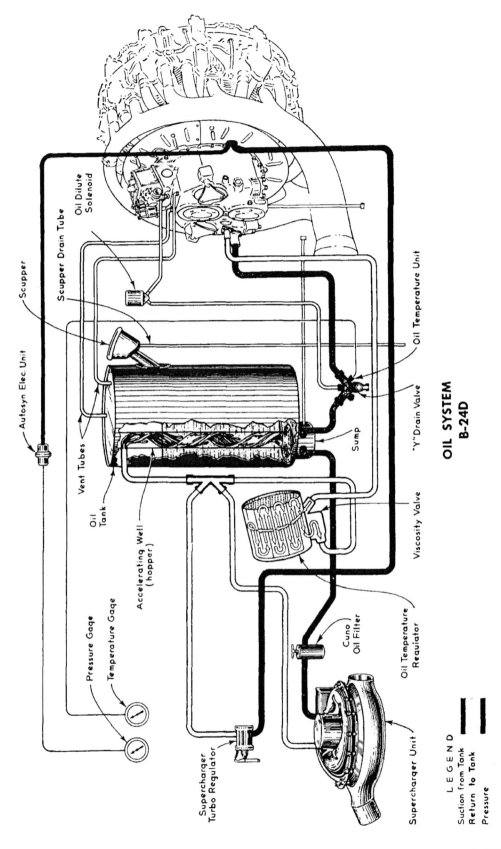

OIL SYSTEM B-24D

OIL SYSTEM

One thirty-nine gallon metal tank, or thirty-three gallon self-sealing tank, is located in each engine nacelle. Self-sealing tank installation starts with Serial 41-23920 and will be installed retroactive to Serial 41-23640. The system is provided with an automatic temperature control which routes the return oil through a cooler when the temperature exceeds 60° C. (140° F.).

A hopper in each tank partially isolates a relatively small quantity of oil to hasten "warm-up." Return oil descends the hopper spirally to minimize foaming of the oil.

The oil has four functions: First, to lubricate the engine; second, to lubricate the turbo supercharger; third, to actuate the turbo supercharger regulator; and fourth, to operate the hydromatic control and feathering of the propeller.

Oil Dilution: A further means of insuring a quick engine starting and warm-up is provided by diluting the oil with small quantities of fuel from the carburetor for 3 or 4 minutes before stopping engine. This process is controlled by a Pilot-controlled solenoid in each nacelle.

Oil dilution is necessary whenever there is a possibility of the engine temperatures dropping below 5° C. (41° F.).

At the end of each engine run, when this possibility exists, the following procedure should be adhered to (this applies to each engine individually):

If the engine temperatures exceed 50° C. (122° F.) the engines must be stopped and allowed to cool before dilution is attempted, the best dilution temperature is 40° C. (104° F.).

Two engine systems may be diluted simultaneously by means of two double throw switches on the Co-Pilot's Instrument Panel.

Maintain the oil temperature of each engine between 5° C. (41° F.) and 50° C. (122° F.), by running the engines at approximately 800 RPM. Speed in excess of 800 RPM will result in oil temperatures higher than 50° C. (122° F.). In this case, the fuel will vaporize and leave the oil with its original viscosity. This condition will also create a fire hazard, resulting from fuel vapors exhausting from the breather outlets.

NOTE: Ideal oil temperature when diluting is 40° C. (104° F.).

Fuel booster pumps need not run during the dilution period. Hold the oil dilution switch in the "ON" position for 4 minutes. If the dilution period is less than 4 minutes, inadequate thinning of the oil will be the result. During the dilution period, the pressure recorded on the fuel pressure gauges should drop from a normal reading of approximately 15 p.s.i. to about 2 p.s.i. If this pressure drop does not occur, check the oil dilution electrical circuits, dilution valves and lines, or fuel pressure gauges for the source of the trouble. Continue to hold the dilution switch in the "ON" position until the engine stops rotating, to insure complete oil dilution. When the engines are started again, subsequent to dilution, a strictly normal procedure should be followed.

If, upon starting engines, (following dilution of the oil) the fuel pressure gauges do not immediately record a normal pressure, check the dilution solenoids for being open.

Oil System Indicators: (See photo Page 2).

Oil Temperature—Gauges on the Co-Pilot's Instrument Panel measure the oil temperature via a thermocouple in the "Y" drain valve.

Oil Pressure—Gauges on the Co-Pilot's Instrument Panel measure oil pressure in the crank case.

Location of Controls:

Oil system controls which affect the propeller are shown on Pages 2 and 4.

Oil Dilution Switches   On Co-Pilot's Switch Panel

See also "Location of Controls," **Page 94.**

ELECTRICAL SYSTEM

ELECTRIC SUPPLY

Power Supply:

1. When the engines are operating, power is generated by four 24 volt, 200 ampere, type P-1 generators, one on each engine, with a total capacity of 22.4 kilowatts. The voltage of each generator may be adjusted at the voltage regulator under the flight deck on each side of the centerline.

2. Main battery power is supplied by two 24 volt, 34 ampere-hour batteries wired in parallel.

3. Auxiliary power is supplied by a type C-10 auxiliary generator, with a capacity of 2.0 kilowatts and powered by an independent gasoline engine ("Homelite" unit). This auxiliary generator must be run for starting engines, or, in the case of main generator failure in flight. The auxiliary engine is not supercharged and power generation from the auxiliary unit, therefore, ceases at high altitudes.

4. For ground operation a provision is made for an external (battery cart) connection. Always use battery cart for first starts, where available, or have auxiliary unit in operation. The excessive loads incident to initial start will shorten the life of the main batteries.

    NOTE: The battery switches must be left "OFF" when using battery cart.

Electrical Systems:

1. 24 volt D.C. single wire system. Most of the electrical equipment in the airplane is supplied through this system.

2. 26 volt A.C. system for the autosyn indicators.

3. 115 volt A.C. system for the fluorescent lighting and the radio compass. Two independent inverters controlled by a selector switch on the Pilot's Pedestal permit use of either unit.

4. 3 volt A.C. system for compass lighting.

5. Miscellaneous systems for the gun turrets, Automatic Flight Controls and radio.

Fuse Boxes and Circuits: From the various fuse boxes to which the above power is delivered, the following sixteen D.C. and A.C. primary circuits distribute power to the mechanisms those circuits operate:

*Code letter of circuit, used on Consair wiring diagrams in aircraft data case at left of Pilot.

Name of Circuit	
Interior Lights	L
Exterior Lighting	E
Heating and Ventilating Controls	A
Bomb Release and Signals	B
Propeller Controls	C
Ice Elimination Controls, Fuel and Hydraulic Pumps	D
Instruments	G
Ignition	H
High Tension	HT
Automatic Flight Controls and Turrets	FL or K
Landing Gear Signals	M
Power, Main	P
Radio and Communication	R
Engine Starter	S
Engine Controls	T
Misc. (Camera, Alarm Bell, etc.)	V

Fusible links (limiters) are added at the four generators, at the batteries, and at several vital points along the power circuit. These links function much like fuses but are calibrated to withstand higher initial currents.

Of the sixteen primary circuits and power circuit, two are described below. The others are described in connection with the systems they control. For example: the Ignition Circuit is under "Power Plant," etc.

Interior Lighting:
1. Lights and easily accessible switches are located throughout the airplane. There are also five extension lights, as follows:
    a. In Bombardier's Compartment.
    b. Forward of Station 1.0 left side on A.F.C. mechanism.
    c. In Radio Compartment over bomb bay.
    d. In lower turret.
    e. In tail turret.
2. Fluorescent light, for the Pilot's Compartment only, comes from four lamps in the instrument panel area.

Exterior Lights: There are five circuits of exterior lights controlled by switches in the Pilot's Compartment.

1. Landing Lights—One retractable landing light is located in the bottom surface of each wing.
2. Passing Light—One **red** passing light is located in the left wing leading edge.
3. Formation Lights—There are seven blue formation lights, three on top of the fuselage and four on the top of the horizontal stabilizer.

4. Running Lights—There are six running lights, one on each wing tip, top and bottom, and one on each vertical fin.

5. Recognition Lights—There is one white light on top of fuselage over center of the wing and three lights (red, green, and amber) under the catwalk.

   NOTE: Although recognition lights are **exterior** lights, they are coded as "L" circuits, same as **interior** lights.

Panels and Switchboards:

Generator—Main control panel forward face bulkhead at Station 4.1, left side flight deck. Carries four field switches to cut generators in or out of main system; one voltmeter with multi-point selector switch to show voltage output of each generator or main bus, and the four ammeters, showing current flow, each for one generator.

Voltage regulators, two each side forward of bulkhead at Station 4.1 under flight deck, provide for generator voltage adjustment for balance of load.

Five main electric switch panels control the **distribution of power** to the sixteen primary circuits. One of these is at the left of the Bombardier; the other four are in the Pilot's Compartment.

Spare Fusible Links, Fuses and Lamps: The fusible links for the four main generators are not accessible in flight; neither are the landing light filament circuit fuses.

Fuses and interior fusible links are replaceable in flight and are located as follows:

1. Spare fusible links are located in the limiter boxes which are located as follows: two on left accumulator bracket; four on the left and six on the right rear face bulkhead at Station 4.1. All links require a small wrench to remove and install.

2. Spare fuses are provided in each fuse box.

3. A spare bulb for the landing gear down position indicator is clipped to the instrument panel.

4. A spare bulb assortment is located aft of bulkhead at Station 4.0 on the left side.

No bulbs for exterior lights are carried.

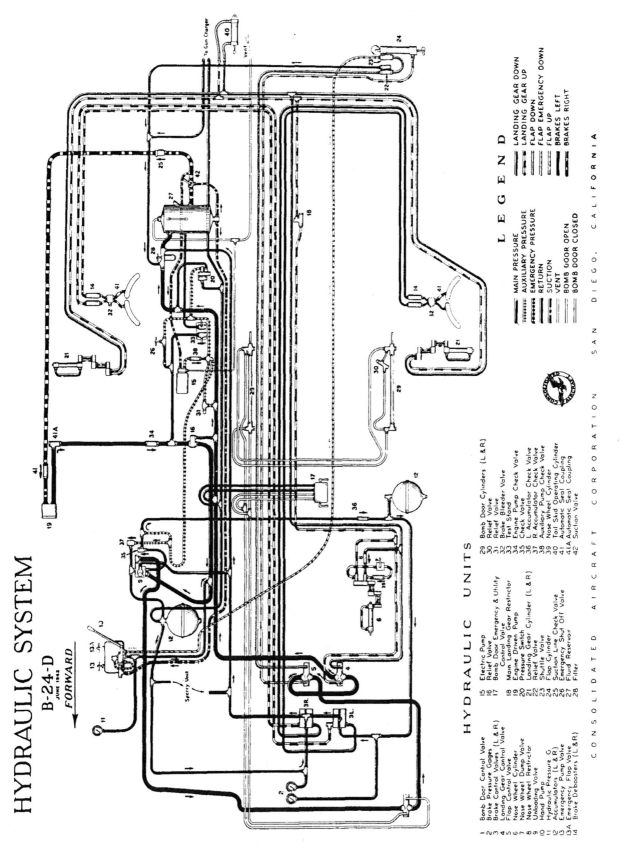

HYDRAULIC SYSTEMS

General:
1. The main hydraulic system operates the tricycle landing gear (including tail skid gear), wing flaps, bomb bay doors, power brake, and the gun-charging mechanism on the belly turret.

2. The hydraulic shock absorber units cushion the landing impact and taxiing loads on the tricycle landing gear.

3. The hydraulic nose wheel shimmy damper unit dampens the tendency of the nose wheel to turn or "shimmy" from side to side.

4. The hydraulic tail turret unit controls the rotation of the rear turret, the elevation of the guns, and the gun charging mechanism.

MAIN HYDRAULIC SYSTEM

General: The hydraulic system consists of a main open center system and a secondary accumulator system.

1. In the open center system the fluid circulates freely in a completely closed circuit when no hydraulic mechanisms are operating. It operates the Bomb Doors, Wing Flaps, and Landing Gear.

2. In the accumulator system the fluid is under constant high pressure, built up in two accumulators. This system is the sole source of brake operating and gun charging pressure, and an auxiliary source of bomb door operating pressure.

HYDRAULIC PUMPS

A Vickers positive displacement pump, driven by No. 3 Engine, supplies pressure for the main system. The pump normally floats on the line. When the flow is diverted to operate an hydraulic mechanism, by closing a valve, pressure builds up to that required to operate the mechanism.

This pump's secondary function is to maintain a charge in accumulator system. An automatic unloading valve in the engine-driven pump pressure line regulates this operation.

An auxiliary electrically-driven pump, located in the right side of the fuselage in the forward bomb bay, maintains accumulator pressure when the main pump is inoperative. An automatic pressure switch and a manual master switch control the pump motor.

95. Fuel Selector, Shut-off, and Cross-over Valve
111. Electric Hydraulic Pump Switch
112. Typical Oxygen Outlet
113. Electric Pump Cross-over Valve
114. Electric Hydraulic Pump

When the engine-driven pump fails, an emergency valve just above and forward of the electric motor may be turned on to connect the auxiliary pump into the main system.

The hydraulic hand pump is located outboard of the Co-Pilot's seat. This pump has a displacement of 2 cubic inches per cycle and can deliver 1000 pounds pressure to the line. It may be used to operate any hydraulic service in an emergency. It operates through a separate line to the wing flaps.

Reserve Fluid: In the event of low fluid, the engine-driven pump and the electrically-driven pump may be connected to the bottom of the reservoir by closing the valve provided in the reservoir outlet. **This should be done only after steps have been taken to insure that no further loss of fluid can take place.**

CAUTION: The landing gear, main bomb door and flap retracting systems cannot and must not be operated simultaneously.

92. Hydraulic Reservoir   114. Electric Hydraulic Pump
117. Hydraulic Suction Line Valve

OPERATING PRESSURES

The main system pressure gauge on the Instrument Panel should indicate approximately 50 pounds with no controls operating. With any system being used, this pressure should rise to between 100 and 1100 pounds.

The wing flaps should be operated before flight to allow the Pilot to check the system and also give the Co-Pilot a check on the operating pressures built up at the gauge. The brake pressure gauge should always show a pressure of approximately 850 to 1000 pounds per square inch.

Location of Controls:

Landing Gear and Tail Bumper Valve Handle	Left side of Pilot's Control Pedestal.
Electric Pump Cross-Over Valve	In forward right bomb bay, at pump. On airplanes equipped for bomb bay tanks—forward of pump near Station 4.1.
Electric Pump Master Switch	Aft of Station 4.1 to right of hatch opening.
Hand Pump	Outboard of Co-Pilot, on floor.
Landing Gear Crank & Gear Box	Centerline of front spar.
Horn Interruption Switch	Pedestal Switch Panel.
Suction Line Valve	Near the reservoir, middle of right side of bomb bay, under wing tanks.
Flap Control Valve	Right side of Pilot's Control Pedestal.
Valves to Main System and Wing	

130. Auxiliary Bomb Door Handle (Inside View)

131. Auxiliary Bomb Door Handle (Outside View)

Flap Lines	Just outboard of hand pump.
Bomb Door Control Valve	Station 0.2, left side of Bombardier's Compartment
Auxiliary Bomb Door Control Valve	Just forward of Station 4.0 at right of hatch opening, and under flight deck floor.
Outside Lock to Auxiliary Bomb Door Control Valve	Just forward of right bomb bay door and under flight deck floor. Special Note: Use main entrance door key.
Pilot's Emergency Pull	On aft end of Pilot's Control Pedestal.
Bomb Door Hand Cranks	Just above bomb bay catwalk at Station 5.0.
Brake Pedals	Hinged to tops of Pilot's and Co-Pilot's Rudder Pedals.
Parking Brake Handle	Aft end of Pilot's Pedestal.

See also "Location of Controls," Page 94.

LANDING GEAR HYDRAULIC SYSTEM B24-D

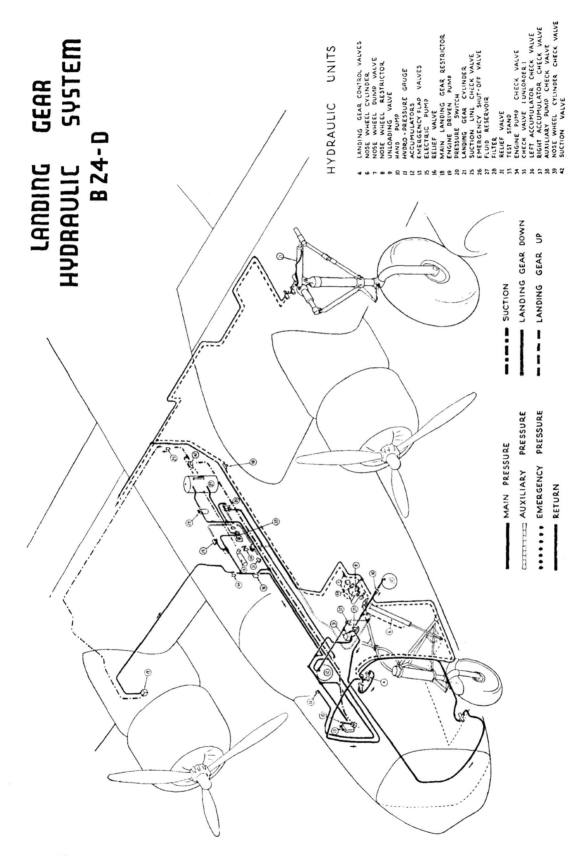

HYDRAULIC UNITS

4. LANDING GEAR CONTROL VALVES
6. NOSE WHEEL CYLINDER
7. NOSE WHEEL DUMP VALVE
8. NOSE WHEEL RESTRICTOR
9. UNLOADING VALVE
10. HAND PUMP
11. HYDRO-PRESSURE GAUGE
12. ACCUMULATORS
13. EMERGENCY FLAP VALVES
15. ELECTRIC PUMP
16. RELIEF VALVE
18. MAIN LANDING GEAR RESTRICTOR
19. ENGINE DRIVEN PUMP
20. PRESSURE SWITCH
21. LANDING GEAR CYLINDER
25. SUCTION LINE CHECK VALVE
26. EMERGENCY SHUT-OFF VALVE
27. FLUID RESERVOIR
28. FILTER
31. RELIEF VALVE
33. TEST STAND
34. ENGINE PUMP CHECK VALVE
35. CHECK VALVE (UNLOADER)
36. LEFT ACCUMULATOR CHECK VALVE
37. RIGHT ACCUMULATOR CHECK VALVE
38. AUXILIARY PUMP CHECK VALVE
39. NOSE WHEEL CYLINDER CHECK VALVE
42. SUCTION VALVE

— MAIN PRESSURE
▦ AUXILIARY PRESSURE
••••• EMERGENCY PRESSURE
— RETURN
--- SUCTION
— — LANDING GEAR DOWN
— - — LANDING GEAR UP

HYDRAULIC SYSTEMS

LANDING GEAR AND TAIL BUMPER HYDRAULIC SYSTEM

General—The landing gear, two main wheels, nose wheel, and the tail bumper gear are simultaneously operated under hydraulic control: The main control for extending and retracting the gear is located on the left side of the Pilot's Pedestal. Movement of the operating lever is restrained by an electric solenoid which is controlled by two switches in series. One switch is push button in the operating handle itself, the other a "micro switch" located on the left landing gear fairing. Extension of the landing gear strut after take-off, closes the "micro switch" and allows the circuit to be completed by pressing the button on the valve operating lever. The solenoid which latches the lever is located forward of the lever on the back of the pedestal, and restrains lever from "UP" position ONLY.

Movement of the selector valve to the "UP" position applies hydraulic pressure simultaneously to the side gear restrictor and to the nose wheel actuating cylinder. The side gear restrictor, restricts the flow of fluid to the main landing gear until the pressure reaches 800 p.s.i. This pressure is sufficient to house the nose gear. When pressure exceeds 800 p.s.i. the restrictor opens and allows fluid to go to the side gear cylinders.

On the lowering operation, pressure is applied to all three gear cylinders simultaneously.

In case of insufficient pressure in the hydraulic system, the hand pump may be used.

In case of complete failure of the hydraulic system, the tricycle landing gear may be manually lowered. See "Emergency" section, Page **86**. No means of manual control is provided for the tail bumper gear.

MAIN LANDING GEAR

Each main landing gear mechanism operated by the main retracting cylinders through "over-rides" is equipped with two latches.

When the main gear is fully extended, a spring-loaded latch on the side brace knee holds the side brace rigid and locks the gear in place. Another latch on the side brace pivot in the wing, locks the gear in the retracted position.

The main gear down latch is painted yellow and can be seen for "down latch" check from the side window. It cannot be seen if flaps are lowered.

NOSE WHEEL GEAR

The nose wheel retracts into the nose of the fuselage under the Pilot's floor. The nose wheel doors are mechanically connected to the gear mechanism so that they open automatically before the gear is extended and close after the gear is retracted.

The nose wheel is designed to caster 45° for free ground maneuverability. An hydraulic shimmy damper tends to restrain any oscillation of the gear about its vertical axis. An internal centering cam in the oleo returns the wheel to its straight ahead position when the oleo is fully extended. A single latch on the drag link actuated by the hydraulic jack override, locks the nose wheel gear in both the retracted and extended positions.

TAIL SKID AND TAIL BUMPER GEAR

A retractable tail skid and bumper is installed on airplanes beginning with 41-23640. It may be used within certain limits on tail low landings. Do not land skid first.

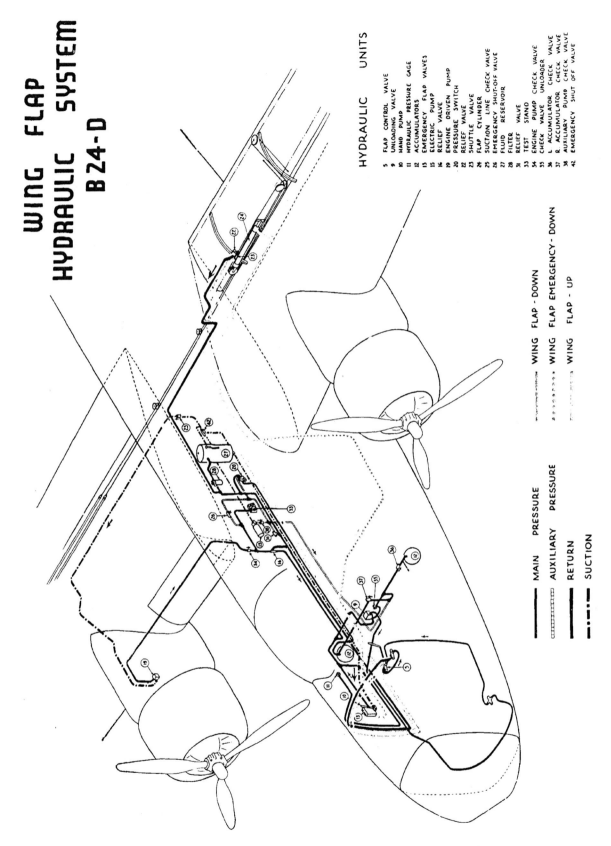

The tail bumper protects the bottom of the fuselage in case the airplane should accidentally tilt back.

WARNING SIGNAL AND LIGHT

A green light on Pilot's Instrument Panel is lighted whenever the landing gear is down and locked.

Further warning that the gear has not been extended is given by an electric horn connected to the throttle controls. When the throttles are moved backward to approximately three-fourths closed, and all landing wheels are not extended and locked, the horn will blow until the gear has been extended and locked or until the throttles are opened to higher engine speed. The horn may be silenced by pressing the Pilot's Interruption Switch on the Pilot's Electrical Switch Board. The horn will then remain silent until the throttles are moved again. This re-sets the horn relay so that another closing of the throttles would again actuate the horn. The horn interruption switch is provided in the event it is necessary to continue flight with one or more engines throttled.

On ships equipped with bottom turrets, this warning is also given when the turret has not been fully retracted.

> NOTE: The green light indicator is wired through switches on all three landing gear units. On ships prior to 41-23750 the warning horn is wired through switches on the left main gear and the bottom turret. Starting with ship 41-23750, the horn is wired through switches on all three landing gear units.

WING FLAP HYDRAULIC SYSTEM

General—The Fowler type wing flaps are operated by a single hydraulic jack which lies along the left rear wing spar at Wing Station 3.0. The flaps move along tracks in the trailing edge and are extended and retracted by a lever on the right side of the Pilot's Pedestal. To raise flaps, move lever forward; to lower flaps, pull lever aft.

In the flap down position, excessive speeds in excess of 155 MPH will create a sufficient pressure on the flaps to open a relief valve at the operating cylinder and allow the flaps to retract automatically.

> CAUTION: This relief valve is a safety precaution only. Do not test during flight as the excessive pressures required for this operation might damage the mechanism.

In case of partial failure of the main hydraulic system, the hand pump may be used through an independent direct line to the flap cylinder to EXTEND ONLY.

To Operate: The hand pump valve to the main system (forward valve) is normally safetied "open;" turn "OFF" and turn the rear valve to the independent flap line "ON."

> CAUTION: After using the hand pump to lower the flaps, always turn the forward valve on to relieve the pressure and allow the shuttle to return to normal. Return both valves to their original position.

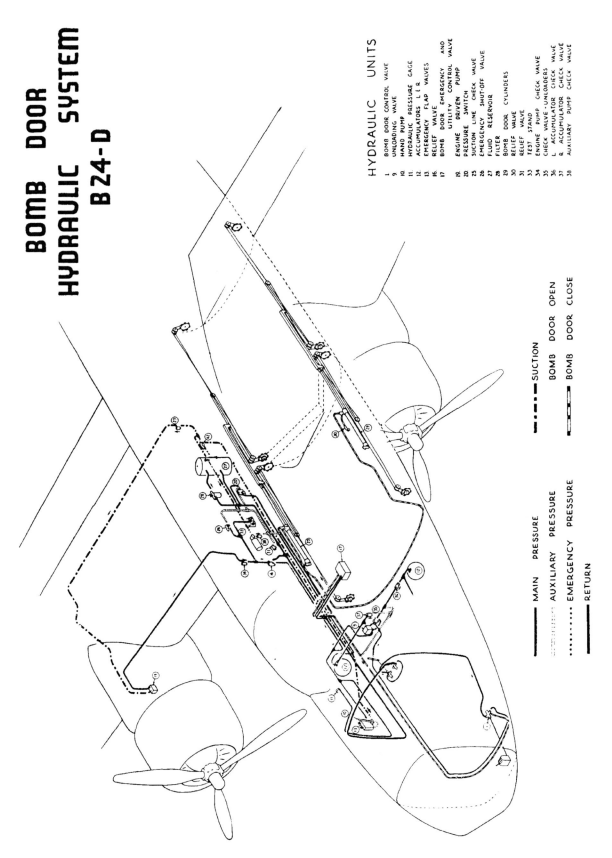

In case of complete failure of the hydraulic system, no manually controlled system is provided for the wing flaps.

Flap Indicating System. See photo on Page 2.

A Selsyn type indicator on the Pilot's Panel shows the flap position at all times.

BOMB BAY DOORS' HYDRAULIC SYSTEM

Each side (2 units) of the bomb bay doors is actuated by an individual hydraulic jack.

The system is hydraulically controlled from any one of four positions:

1. Bombardier's Compartment — Main control valve
2. Under Radio Operator's floor at hatch opening. — Auxiliary control valve
3. On the ground from access door on right side forward of bomb door. — Auxiliary control valve
4. Pilot's Compartment — Emergency operation of auxiliary valve. Doors may be opened but not closed until pull line is reset.

CAUTION: The Pilot's emergency pull line to the auxiliary valve cam (see No. 4 control in above paragraph) must be re-set by hand or hydraulic system will by-pass through the bomb jack relief valve thus affecting the entire hydraulic system.

Under military operating conditions the main control valve is used to control the operation of the doors.

The auxiliary valve, in the accumulator system, is generally used for local flight operations.

In case of complete failure of the hydraulic system, the doors may be manually operated by hand cranks accessible from the catwalk at the center of the bomb bay.

Bomb Bay Door Indicating System:

When these doors are fully open the following lights are illuminated:

1. A red light on the Bombardier's Panel.
2. An amber light on the Pilot's Panel (see photo, Page 2).
3. A white light on the tail to notify other airplanes in the formation.

POWER BRAKE HYDRAULIC SYSTEM

Two completely separate units operate the hydraulic brakes. Each unit contains two brake cylinders which control one of the dual Hayes expanding bladder type brakes on each main landing wheel. One cylinder of each unit is mechanically interconnected to the right hand brake pedal of both Pilot and Co-Pilot; the other cylinder of each unit is similarly connected to both left hand brake pedals.

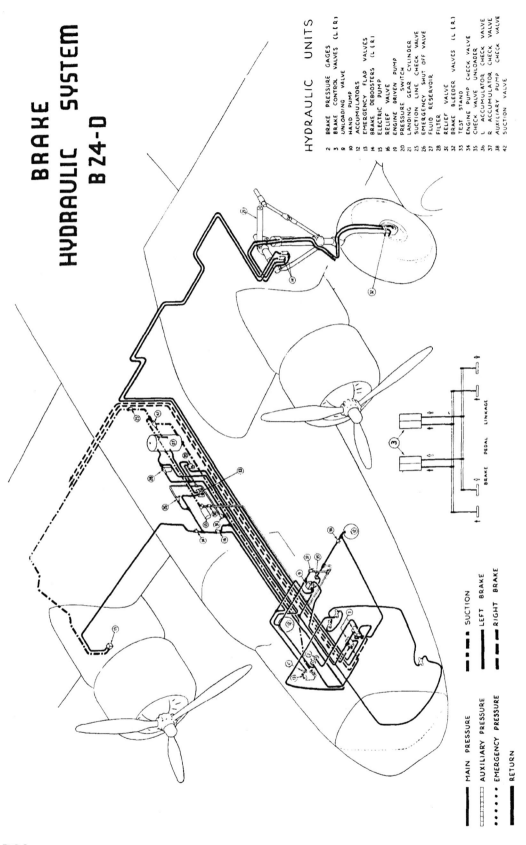

Each unit takes its pressure directly from a different one of the two main accumulators which are isolated from each other by check valves so that failure of one accumulator does not affect the other. Failure of one complete unit leaves one-half braking power available.

Parking Brake:

To Set: A lever on the left rear corner of the Pilot's Pedestal is raised to engage locking cam while brakes are held on.

To Unlock: Press brake pedals.

Brake Indicating System (See photo, Page 2).

Two pressure gauges on the instrument panel indicate the pressure in each brake system.

NOSE WHEEL SHIMMY DAMPER HYDRAULIC SYSTEM

An independent system, located on the nose gear strut, dampens out nose wheel shimmy by means of an accumulator connected to two hydraulic damper rams which act in opposite directions. A pressure gauge is attached to the accumulator. The pressure must be maintained at 150 to 250 p.s.i. When the pressure falls below 150 each of the two cam slots in the damper may be chocked with a piece of metal.

CAUTION: This emergency "jamming" will eliminate the wheel vibration, but will also fix the nose wheel in straight travel, non-steerable. Therefore, this procedure must only be used in an emergency as it will injure the wheel and damper mechanisms.

NOTE: An emergency positive type nose wheel lock is to be provided in the near future. They will be furnished for all delivered airplanes.

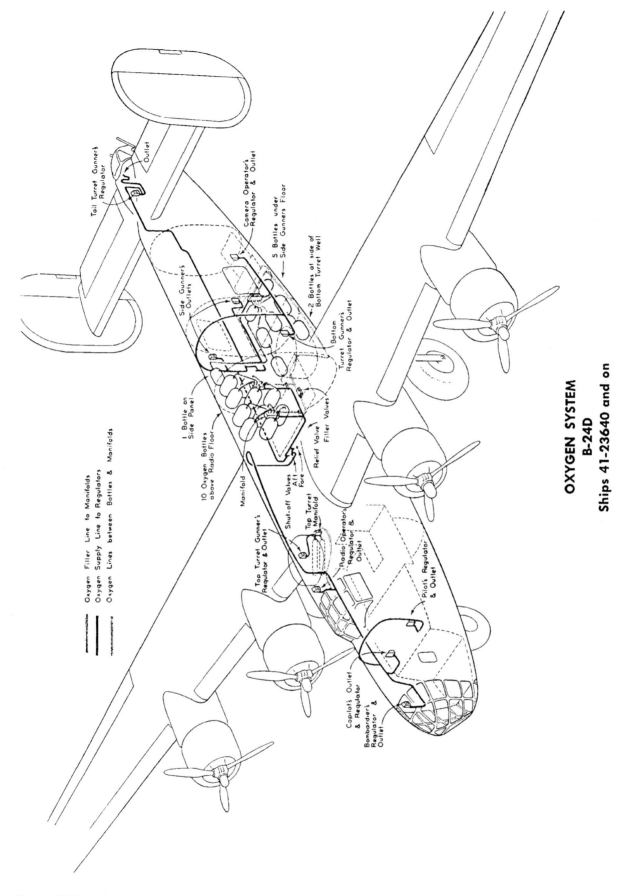

OXYGEN SYSTEM

A shut-off valve on the rear spar permits shutting off the oxygen supply to all outlets. This valve is normally safetied open. A manually operated flow regulator is at each outlet.

Oxygen is supplied by 10 to 18 oxygen bottles located at various points in the fuselage aft of the flight deck. Airplanes not equipped with auxiliary wing tank fuel cells have only 10 bottles (5 in each wing outboard of wheel wells).

Oxygen Indicators—Regulator dial at each outlet is marked in thousands of feet. When the regulator is set at the flying altitude, an attached flow dial indicates amount of flow.

Location of Controls

Oxygen Outlets   At each crew station and at right of flight deck hatch in bomb bay.

Main Shut-Off Valve   On rear spar to right of center line. Available from radio compartment over rear bomb bay.

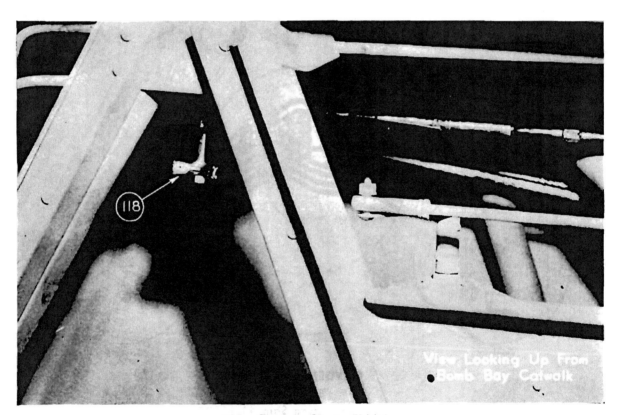

118. Oxygen Shut-off Valve

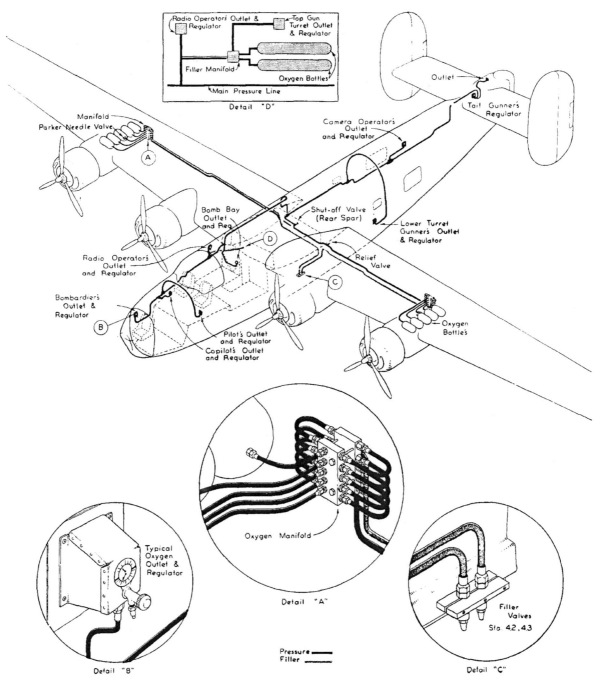

**OXYGEN SYSTEM
B-24D
Up to and including Ship 41-11938**

CONSOLIDATED AIRCRAFT CORPORATION
San Diego, California

MAN HOUR OXYGEN CONSUMPTION TABLE

In using this table it is well to remember that the figures given are averages and that the demand for oxygen varies with the individual. No attempt, therefore, is made here to outline specific procedure. Generally, however, oxygen is required in flight from ten to twelve thousand feet after one hour; above fifteen thousand feet continuous use of oxygen is necessary.

Altitude in Feet	Volume Consumed Per Man Hour		DURATION OF OXYGEN IN HOURS														
			10 Bottle System Total Capacity—287.8 cu. ft. *No. of men in crew					18 Bottle System Total Capacity—518.04 cu.ft. *No. of men in crew					20 Bottle System Total Capacity—575.6 cu. ft. *No. of men in crew				
	Cu. ft.	Liters	6	7	8	9	10	6	7	8	9	10	6	7	8	9	10
10,000	2.68	76	17	15	13	11	10	32	27	24	21	19	35	30	26	23	21
15,000	3.67	104	13	11	9	8	7	23	20	17	15	14	26	22	19	17	15
20,000	4.66	132	10	8	7	6	6	18	15	13	12	11	20	17	15	13	12
25,000	5.72	162	8	7	6	5	5	15	12	11	10	9	16	14	12	11	10
30,000	6.78	192	7	6	5	4	4	12	10	9	8	7	14	12	10	9	8

* This man-hour demand is applicable to the common system only, and does not include the Top Gunner whose separate equipment consists of two D-2 Bottles with a combined capacity of 13.78 cu. ft.

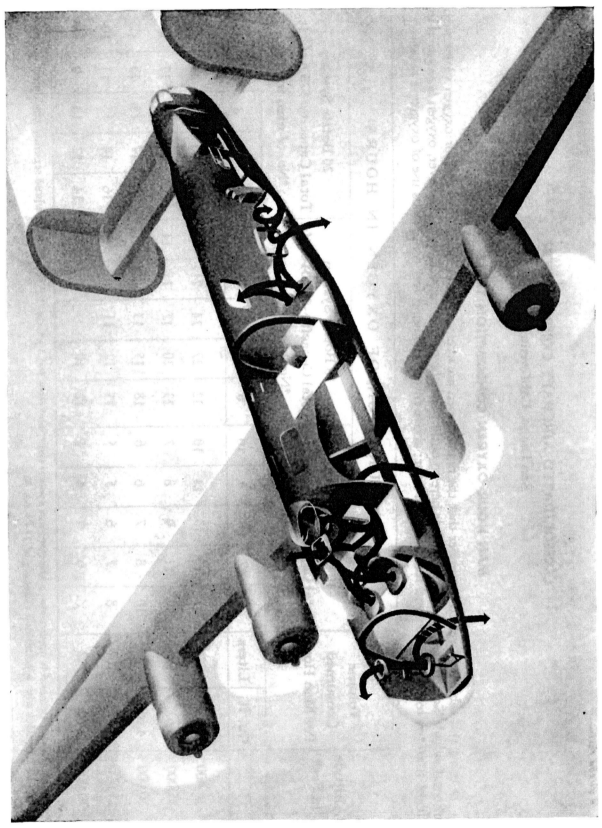

EMERGENCY EXITS B-24D

EMERGENCY EXITS AND EQUIPMENT

Five entrances and exits are provided:

1. The main entrance door into rear compartment—This should always be locked when leaving the airplane unattended; it should always be unlocked before take-off. To open from the outside: unlock and push up. To open from the inside: pull handle and lift.

2. The bomb bay doors into the bomb bay—All four doors operate simultaneously under hydraulic pressure, and in pairs on each side operated manually.

   To open from the outside, unlock a small door in the skin just forward of the bomb bays on the right side of the airplane. Pull outboard on valve handle extension to open doors; push inboard to close doors. A second operating handle of the same valve is located on the right side ahead of the bomb bay under Pilot's Deck on right side.

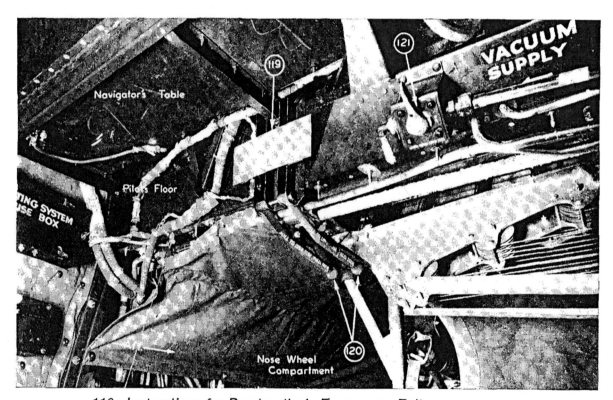

119. Instructions for Bombardier's Emergency Exit
120. Nose Wheel Pull Handles for Bombardier's Emergency Exit
121. Vacuum Valve for Nose Compartment Camera

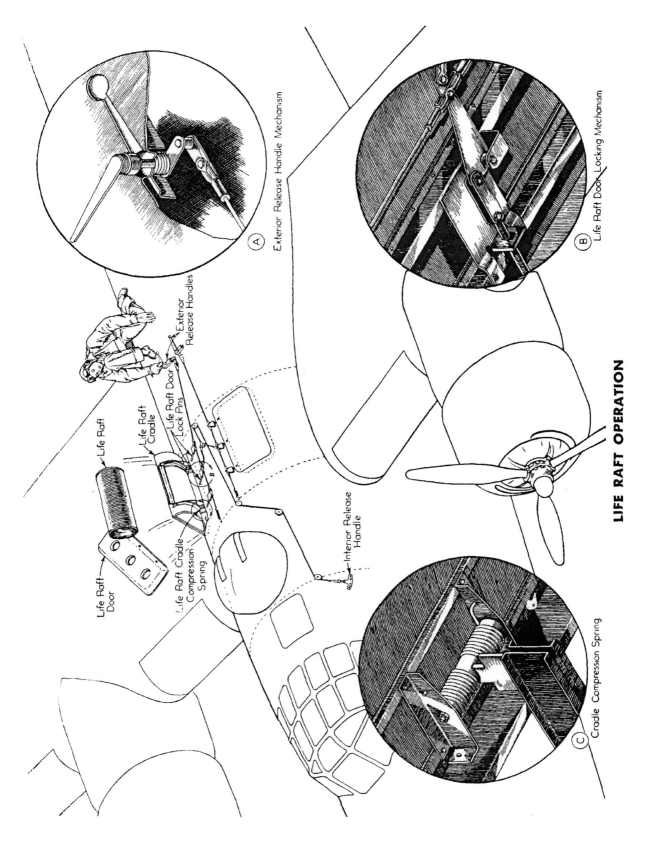

LIFE RAFT OPERATION

Page 84] EMERGENCY EXITS AND EQUIPMENT

RESTRICTED

In an emergency only, the Pilot can open but cannot close this valve by means of a pull handle on the control pedestal. After such an emergency operation, the pull handle line to the valve must be reset by hand.

The doors may be operated manually with the control valve in "OPEN" position by turning the cranks located above the catwalk at Station 5.0.

CAUTION: Either the main valve in the Bombardier's Compartment or the auxiliary valve must be in "OPEN" position to operate.

3. The nose wheel doors—In an emergency, each door is operated by a cable connected to a lever indicated by an instruction plate just aft of the Bombardier. To open: pull lever down as far as it will travel.

4. The escape hatch—Just aft of the Pilot, opens inward by pulling the handle down.

5. The two side doors—They are unlatched from either the inside or the outside and swing in and up.

EMERGENCY EQUIPMENT

LIFE RAFTS—Two Type A-2 life rafts are carried in the fuselage above the wing between Stations 4.2 and 4.4. To release either raft from inside the airplane, pull the "T" handle at the center of the airplane on the upper part of the forward face of bulkhead at Station 4.0. On B-24D No. 41-23640 and on, the "T" handle is located immediately aft of the top escape hatch. The pull cable releases the lock pins holding the life raft doors closed and allows the spring bungee to throw the raft out, clear of the fuselage. A rip cord attached to the raft cradle automatically opens the valve which controls the raft inflation from the CO^2 bottle. To release either raft from outside the airplane, the lever flush in the fuselage aft of each door should be lifted up and twisted 90 degrees. This action pulls the same cable that attaches to the "T" handle on the inside and releases the raft in the same manner as described above. **Do not release rafts until plane is at rest in the water.**

FLARES—Flare ejector tube is located on the left of the center line of the airplane immediately forward of entrance door, between Stations 7.2 and 7.3. Flares are stowed outboard.

To Load a Flare: Move operating handle downward. This rotates cam so that flare can enter tube. Insert the flare to the proper position where cam enters the slot in the side of the flare casing. Connect the flare safety to the fish line on reel.

To Eject a Flare: The flare tube is fitted with two controls: one a toggle handle located on the inboard side of the tube which opens the door in the bottom of the fuselage uncovering the flare tube, and the second handle, located on the aft side of the flare tube, which is the flare handle.

Pyrotechnics: The pyrotechnic installation located on the left side of flight deck between Stations 3.0 and 4.0 consists of:

1 Type M-2 Signal Pistol	9 Type M-10 oM-11 Signals
1 Type A-1 Portable Signal Container	1 Type A-1 Holder, Pyrotechnic Pistol

On B-24D airplanes No. 41-23640 and on, stowage has been changed to rear compartment right side between Stations 7.4 and 7.5.

PARACHUTES—Parachutes are stowed on racks convenient to all crew members.

LANDING GEAR
EMERGENCY OPERATION

Main and Nose: The manual emergency system is used to lower and lock—IT WILL NOT RAISE—the main and nose landing gear in case of total failure of the main, auxiliary (electric) and emergency (hand) hydraulic systems. The operating crank is located on the forward center section spar on the centerline and can be reached from the forward end of

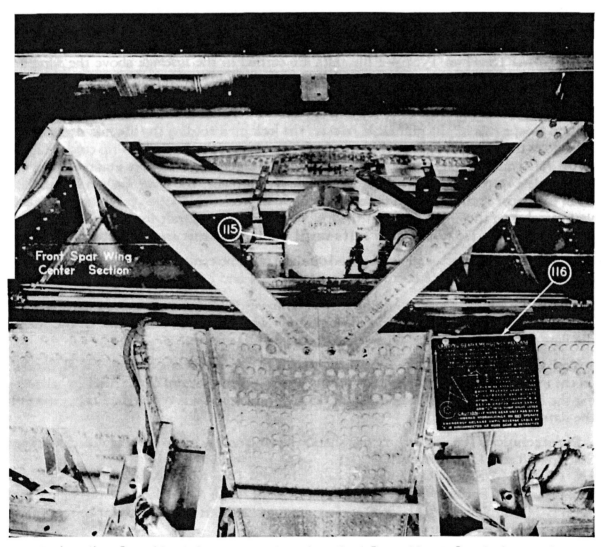

115. Landing Gear Hand Crank   116. Landing Gear Hand Crank Instructions

the catwalk. The nose and main gear are operated simultaneously by the crank. See "Functional Operations" below.

TO LOWER:

1. Place main operating or selector valve in the "DOWN" osition and <u>hold it so until</u> the operation is completed.

2. Rotate the hand crank clockwise or to the right facing aft; 30 turns are required to complete the operation. The safety wire is broken and the gear "UP" latches are released by the first few turns. The pressure on handle, up to approximately 10 turns, reaches about 30 pounds; decreasing after 16 turns it again increases after the 28th turn, performing in succession the operation of unlatching all gear, pulling nose gear over its high point and latching.

3. The main gear will immediately fall by gravity to the "DOWN" <u>unlatched</u> position.

4. The nose gear will fall by gravity after it has been lifted over the high point by the hand crank.

5. Nose wheel doors open mechanically with the first down movement of the nose gear.

6. The last few turns of the hand crank complete the operation and lock the down latches on main gear. Nose gear locks automatically when it falls.

CAUTION: Check down latch "engaged" on nose gear visually from nose wheel compartment. If latch is not engaged push latch into "engaged" position by hand.

Check main gear down latch visually from rear side windows. Latch is engaged when entire yellow tip is visible. A "gap" in the yellow means it is not completely locked. THE LATCHES CANNOT BE SEEN WITH THE FLAPS DOWN. If not completely locked, turn handle further to latch. A hand operated emergency mechanism trips a latch on the nose wheel gear so that the lock will operate. This latch must be reset before the gear is raised hydraulically.

The hand operated mechanism trips the nose gear dump valve which by-passes the "up" side to the "down" side of the landing gear system. This must be reset before the gear is again operated hydraulically.

To reset the emergency crank: rotate the hand crank counter-clockwise 30 turns and reset the trip mechanism on the nose gear.

EMERGENCY NOSE GEAR LOWERING INSTRUCTION

1. Place landing gear lever in down position.
2. Remove curtains around aft end of nose wheel enclosure.
3. Remove latch linkage bolt (pin on later airplanes) at "A."
4. Pry open latch at "B" with fingers (latch shown disengaged).
5. Sit under flight deck as shown and place right foot on shimmy dampener collar at "C."
6. Place both hands under the top of the oleo strut at "D."
7. Push with right foot at "C" and lift with both hands at "D" to extend nose gear.
8. After nose gear is in down position push up on drag strut at "E" and press latch "B" into the gear locked position.

   NOTE: Replace the latch linkage bolt at "A" before normal nose landing gear retraction is attempted.

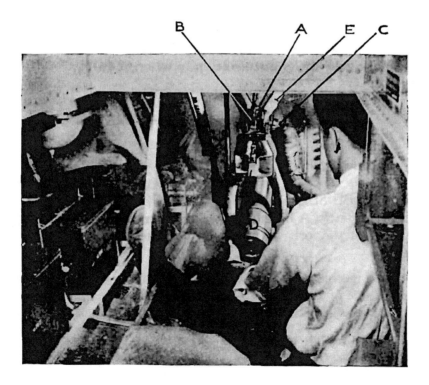

EMERGENCY FLAP OPERATION

In case of failure of the main or auxiliary systems the flaps may be operated by the hydraulic hand pump, located on the right side of the Co-Pilot's seat.

This hand pump is connected to two globe valves located immediately forward of the pump. Normally, the forward one of these two valves is safety wired in the open position so that the hand pump is always connected with the main system through this valve.

To Operate Flaps: Refer to the Hydraulic Chart, **Page** 72.

To operate flaps, move the main operating lever to the "DOWN" position, close the forward valve and open the rear valve which connects the hand pump to the emergency flap hydraulic line leading directly to the extending side of the flap cylinder. Operation of the hand pump then extends the flaps.

When using the hand system the shuttle valve at the flap operating cylinder is moved so as to close off the main system. For this reason before the flaps can be moved in any manner to the "UP" position, the shuttle valve must be returned to the normal position. This is accomplished by opening the forward of the two emergency valves while the rear one is still open.

> WARNING: The flaps cannot be moved until the shuttle valve is returned to normal as an hydraulic lock exists when shuttle valve is operated by the hand pump. Refer to the Hydraulic Chart, **Page** 72 and instructions Page 73.

EMERGENCY BOMB DOOR

The bomb door is normally operated hydraulically from either the main or auxiliary system by the valves and in a manner noted under "Bomb Doors," **Page 75.**

In case of failure of the hydraulic system the bomb doors may be opened or closed manually by two hand cranks, located on each side of the catwalk on Station 5.0 in the center of the bomb bay. These cranks operate a drum upon which a cable is wound, to operate the main bomb door cable crosshead in the same manner as the hydraulic piston.

> CAUTION: Either the main valve in the Bombardier's Compartment or the auxiliary valve must be in "OPEN" position to operate.

The bomb door utility valve may be operated manually at the valve, or from the outside of the ship through a door which is normally locked. It may also be operated (Emergency open only) from the pilot's cockpit by a pull handle on the control pedestal. This emergency pull also operates the bomb release mechanism and jettisons (drops) the entire bomb load "unarmed" (safe). The toggle or pull lever is connected to a sliding cam by flexible cables. A roller in the cam slot operates a bell crank connected to the bomb door utility valve by a flexible cable. As the roller slides in the cam slot, it moves the bell crank through the cycle, "Valve Neutral," to "Bomb Doors Open," to "Valve Neutral." In moving to the last position the bomb release mechanism is operated. A safety lock prevents the toggle from being pulled all the way to release the bomb load until the bomb doors are fully open.

> NOTE: The cam must be reset before ship is again loaded with bombs. Page 91.

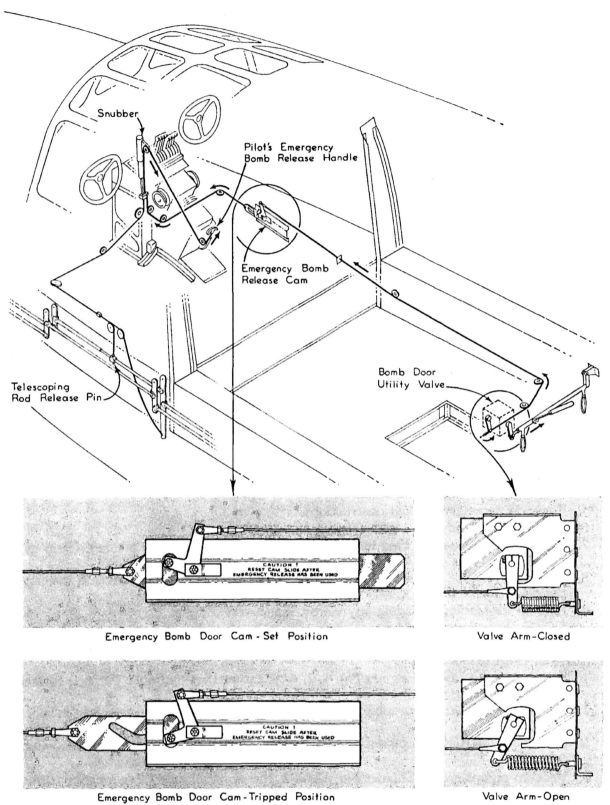

**EMERGENCY BOMB RELEASE SYSTEM
B-24D**

ENGINE FAILURE ON TAKE-OFF

When a motor fails, particularly on take-off, the action to be taken by the Pilot must vary somewhat depending upon the situation. (Such as terrain, weather, traffic, load, etc.) Much depends on the judgment and discretion of the Pilot. The value of an engine failure procedure lies in the fact that it prepares the Pilot in advance of an emergency as to what must be done. The execution of these items can then be done smoothly and efficiently thus eliminating the need for hurried, unpremeditated moves which constitute the greatest hazard of any emergency operation.

There is sufficient power available in three engines to climb the B-24 and all of this power should be used. When sufficient altitude has been gained to insure safety, reduce power to amount required to safely sustain flight and proceed in a normal manner.

Failure of Engines No. 1, 2, or 4:
1. Maintain 135 MPH indicated airspeed.
2. Trim ship—rudder first to relieve yaw, aileron second for as good "hands off" condition as possible.
3. Start gear up.
4. Feather dead engine (mixture—"IDLE CUT-OFF;" fuel booster—"OFF;" ignition —"OFF;" cowl flap—"CLOSED").
5. Power as necessary to clear obstructions.
6. Request emergency landing clearance.
7. When gear is up, raise wing flaps by small amounts (3° to 4° at a time). (Ship will handle better if 5° to 10° of flap is retained at low speeds.)
8. Jettison overload: this should not be necessary under most conditions and is left to the discretion of the Pilot.
9. Line up for landing, avoiding any violent maneuvers doing so.
   NOTE: Engineer should be sent aft to ascertain whether necessary to close fuel valve to dead engine (broken fuel line, etc.).

Failure of Engine No. 3:
1. Same as in the case of Engines No. 1, 2, or 4, except that hydraulic pump is located on Engine No. 3.
2. Be certain auxiliary hydraulic pump is on. Star valve on open center system open (engineer).

If already on instruments, or it becomes necessary to go on instruments immediately after take-off:
1. Vacuum pumps are on Engines No. 1 and 2, therefore, selector valve must be turned to live engine to keep flight instruments in operation. (Engineer)
2. If on Automatic Pilot, turn it off and fly manually.

See Page 23 for "Two Engine Failure in Flight."

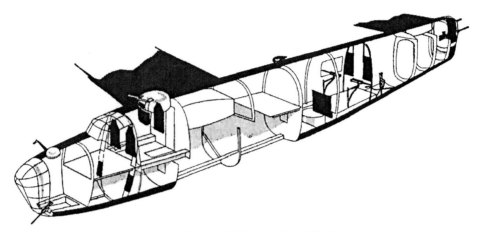

ARMOR PLATE LOCATION

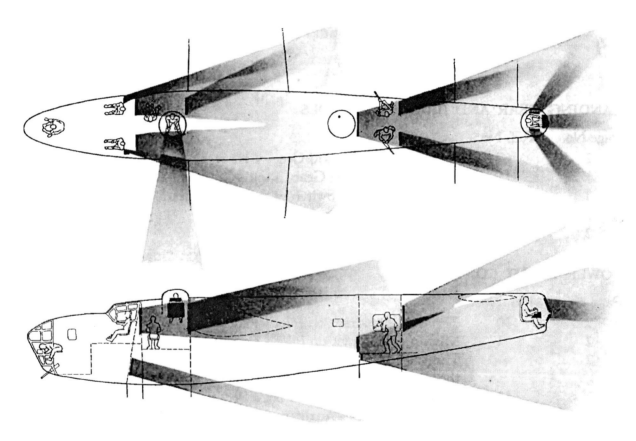

AREA OF ARMOR PROTECTION

CONTROL LOCATION

The location of the majority of the controls listed below are shown on the following photo pages:

FLIGHT CONTROLS:

Page No.	Ref. No.	
2		Aileron—Control Column in Pilot's Compartment
2		Elevator—Control Column in Pilot's Compartment
2	55	Rudder—Rudder pedals in Pilot's Compartment
2	65	Aileron Trim—Aft of Pilot's Pedestal
2	76	Elevator Trim—Left of Pilot's Pedestal
2	61	Rudder Trim—On top of Pilot's Pedestal
2	59	Wing Flaps—Right of Pilot's Pedestal

LANDING GEAR AND BRAKE CONTROLS

Page No.	Ref. No.	
2	67	Landing and Tail Skid Gear—Left rear of Pilot's Pedestal
86	115	Emergency Landing Gear Crank—Centerline of front spar
2	74	Horn Interruption Switch—On Pilot's Pedestal
2	54	Brake Pedals—On top of rudder pedals
2	66	Parking Brake—Aft of Pilot's Pedestal

POWER PLANT CONTROLS

Page No.	Ref. No.	
2	30	Throttle—Quadrant on Pilot's Pedestal
2	34	Mixture—Quadrant on Pilot's Pedestal
2	46	Throttle Lock—Right of above quadrant. (This lock is No. 46 at the **right** of photo)
2	53	Cowl Flap Switches—On Pilot's Pedestal
2	47	Battery Switches—Auxiliary Switch Panel
2	50	Magneto Switches—Ignition Switch Panel
2	41	Starter Switches—Co-Pilot's Switch Panel

POWER PLANT CONTROLS—Continued

Page No.	Ref. No.	
2	128	Starter Crank Stowage
56	8	Generator Switches—Fwd. of Station 4.1
2	71	A.C. Switch—On Pilot's Pedestal

Supercharger

Page No.	Ref. No.	
2	25	Supercharger — Quadrant on Pilot's Pedestal
2	46	Supercharger Lock—Left of above quadrant (This lock is the No. 46 at the left of photo)
2	68	Intercooler Switches — On Pilot's Pedestal

Propeller

Page No.	Ref. No.	
2	78	Propeller Governor Switches —On Pilot's Pedestal
4	79	Feathering Switches — Over windshield

Oil

Page No.	Ref. No.	
2	43	Oil Dilution Switches—Engine Controls Panel

Fuel (Main)

Page No.	Ref. No.	
2	39	Primer Switches—Engine Controls Panel
2	38	Booster Pump Switches—Engine Controls Panel
		Main Shut-off and Cross-over Valves—Under front spar in bomb bay.
53	95	For No. 1 and No. 2 Engines
66	95	For No. 3 and No. 4 Engines
54	93	Transfer Pump—Aft of Radio Compartment over bomb bay
		Vent Disconnect Points—Fwd. of Transfer Pump
		Fuel Gauge Selector Valves—Under gauges, fwd. of Sta. 4.1
		Fuel Gauge Shut-off Valves—Under main fuel cells in bomb bay.
52	87	No. 1
52	89	No. 2
52	90	No. 3
52	92	No. 4
56		Fuel Gauge Vent Valves—Over Gauges, fwd. of Sta. 4.1

128. Starter Crank Stowage

RESTRICTED CONTROL LOCATIONS [Page 95

Fuel (Main)—Continued

Page No.	Ref. No.	
55	129	Fuel Gauge Drain Valves—Low in left side of bomb bay between Station 4.0 and 5.0
52	134	Center Wing Drain Valves—Under main fuel cells in bomb bay.

Fuel (Auxiliary)

Wing System Shut-off and Selector Valve—Aft of transfer pump
Bomb Bay System Shut-off Valves—Catwalk between tanks
Bomb Bay System Booster Pump—Catwalk between tanks

BOMB DOOR CONTROLS:

Page No.	Ref. No.	
98	108	Bombardier's Control Handle—Quadrant to left of Bombardier
68	*130	Auxiliary Control Handle—Just forward to Station 4.0 under flight deck at right of hatch.
69	*131	Outside Auxiliary Control Handle—Just forward of right bomb bay door. Use main entrance door key.

*See "Hydraulic Controls," Page 68.

2	63	Pilot's Emergency Pull—Aft of Pilot's Pedestal
67		Emergency Hand Cranks—Above catwalk at Station 5.0

HYDRAULIC CONTROLS:

The following are general hydraulic controls. They are used in connection with the hydraulic controls of Flight, Bomb Door, Landing Gear, and Brakes.

Page No.	Ref. No.	
66	113	Electric Pump Cross-over Valve—Fwd. right bomb bay, aft of pump. On airplanes equipped with bomb bay tanks, fwd. of pump.
66	111	Electric Pump Switch—Aft of Station 4.1 right of hatch.
67	117	Suction Line Valve—On reservoir, middle of right bomb bay under wing.

ANTI-ICER CONTROLS: (PROPELLER)

Page No.	Ref. No.	
2	37	Rheostat Control—Fwd. of Pilot's Pedestal

ANTI-ICER CONTROLS (WINDSHIELD)

Page No.	Ref. No.	
98		Bombardier's Pump—Fwd. of Station 0.1, right of Bombardier.

DEFROSTING CONTROLS

2	3	Push-Pull Rodd—On Pilot's Panel
2	45	Push-Pull Rod—On Co-Pilot's Panel
98	109	Shutter—On Bombardier's Heater

DE-ICING CONTROLS

Page No.	Ref. No.	
2	52	Control Lever—On Co-Pilot's Panel
56	84	Vacuum Selector Valve—Fwd. face of Station 4.1

EMERGENCY CONTROLS

Page No.	Ref. No.	
86	115	Landing Gear Hand Crank
		Life Raft Release—Aft of Pilot at ceiling
84		Outside Life Raft Release—Top of fuselage, aft of life raft door.
2	33	Demolition Switches
2	73	Alarm Bell Switch—On Pilot's Pedestal
83	120	Nose Wheel Door Pull Handles

HEATING AND VENTILATING

Page No.	Ref. No.	
2	48	Pilot's, Co-Pilot's, and Top Turret—On Auxiliary Switch Panel
		Bombardier, Navigator, and Radio—At Bombardier's Panel
98		Bombardier's Fresh Air Control—Disc on Bombsight Window.
2	10	Pilot's Fresh Air Control—Instrument Panel

INSTRUMENT CONTROLS

Page No.	Ref. No.	
56	84	Vacuum Selector Valve—Fwd. face of Station 4.1
2	60	Pitot Heater—Pedestal Switch Panel
2	71	A.C. Switch—On Pilot's Pedestal
2	70	Fluorescent Light Switch—Pilot's Pedestal

OXYGEN CONTROLS:

Page No.	Ref. No.	
78		Oxygen Outlets—At each crew station and in bomb bay
79	118	Shut-off Valve—Rear spar, right of centerline

RADIO CONTROLS:

See "Radio" Section

RESTRICTED

BOMB CONTROLS:

See "Armament" Manual

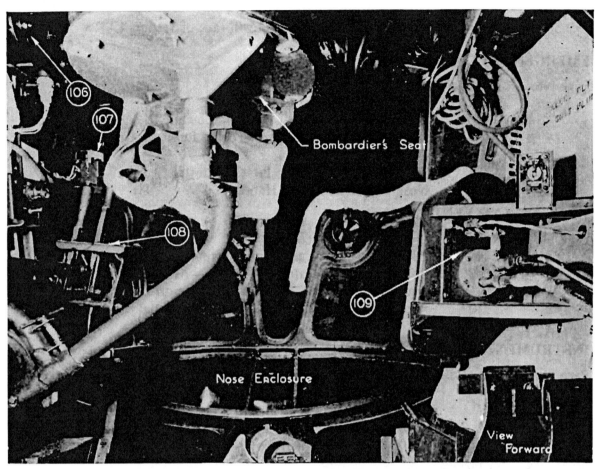

106. Intervalometer
107. Firing Key
108. Bombardier's Quadrant
109. Bombardier's Heater

GUN CONTROLS:

See "Armament" Manual

AUTOMATIC FLIGHT CONTROLS (AFCE)

Page No.	Ref. No.	
2	57	Pilot's Turn Control—Right side of Pedestal
2	4	Blinker Light Switch
2	21	AFC Controls

RADIO EQUIPMENT

All radio equipment used in B-24D Airplanes derives power from the 24V DC supply of the main power system. The radio system includes the following individual systems:

Unit 1   **Interphone system**
Used for intercommunication between crew members and to provide for reception or transmission of radio messages from any crew station except as noted under "Liaison Equipment" later in this volume.

Unit 1-1   One amplifier mounted immediately behind Co-Pilot.

Unit 2   **Command Equipment**
Which is normally used for ship to ship communication.

Unit 2-1, 2-2   2 Transmitters located over wing center section.

2-3, 2-4, 2-5   3 Receivers located over wing center section.

Unit 3   **Liaison Equipment** (of medium power)
Used for ship to base or ship to ground communciation.

Unit 3-1   1 Transmitter on flight deck under Radio Operator's table.
3-2   1 Receiver on flight deck on Radio Operator's table.

Unit 4   **Radio Compass Receiver**
Used for direction finding and in cross-country navigation.

Unit 4-1   1 Receiver located over wing center section right side.

Unit 5   **Marker Beacon**
Used in making instrument landings.

Unit 5-1   1 Receiver located in bomb bay at Station 5.0.

Unit 6   **Filter System**
FL-5C filters **Units 6-3** and **6-4** with BC-345 switch box, for Pilot **Unit 6-1** and Co-Pilot **Unit 6-2**, right and left by Pilot's and Co-Pilot's seats.

RESTRICTED

INTERPHONE EQUIPMENT

The interphone system consists of a type BC-347 Interphone Amplifier, one PE-86 Dynamotor, one BC-366 Jack Box for each crew station, and one T-17 Throat Microphone or T-20 Microphone with RC-19-A Microphone Amplifying Equipment for each crew station. Each throat microphone is equipped with either a CD-318 switch cord or a contractor furnished "push-to-talk" switch as the installation requires. Each jack box is equipped with a headset connector cordage, on one end of which is a PL-55 plug for inserting in the jack box, and on the opposite end, a JK-26 jack into which is plugged the crew member's headset. A stowage hook is provided adjacent to each jack box for stowing the cordages and switch when not in use.

The various interphone stations are located as follows:

Unit		
1-2	Bombardier's	Station 0.1 Right Side
1-3	Pilot's	Station 2.1 Left Side Flight Deck
1-4	Co-Pilot's	Station 2.1 Right Side Flight Deck
1-5	Radio Operator's	Station 3.0 Right Side Flight Deck
1-6	Top Gunner's	Right Seat Support Member of Turret
1-7	Bomb Bay	Station 5.0 Right Cross Member of Bulkhead
1-8	Bottom Turret	Station 6.1 Left Side
1-9	Side Gunner's	Station 7.1—Right and forward of Gun Doors
1-10	Side Gunner's	Station 7.1—Left and forward of Gun Doors
1-11	Camera	Station 7.5 Left Side
1-12	Tail Gunner's	Station 9.1 Left Side

The Bombardier's, Radio Operator's, Top Gunner's, Bomb Bay, Bottom Gunner's, Two Side Gunner's, Camera, and Tail Gunner's Stations are all equipped with standard interphone station equipment which consists of the following:

One BC-366 Jack Box
One T-17 Throat Microphone
One CD-318 Microphone with "push-to-talk" switch and cordage, Units 7-3 and 7-4.
One headset cordage with PL-55 plug and JK-26 jack.

NOTE: The output of the Pilot's or Co-Pilot's throat microphones is connected to the interphone or radio circuits, through push-button switches, Unit 7-1, 7-2 located on the Pilot's or Co-Pilot's Control Wheels, Page 103. These switches are functionally identical with the push-to-talk switches in the CD-318 cordage, and when depressed, give the same operation.

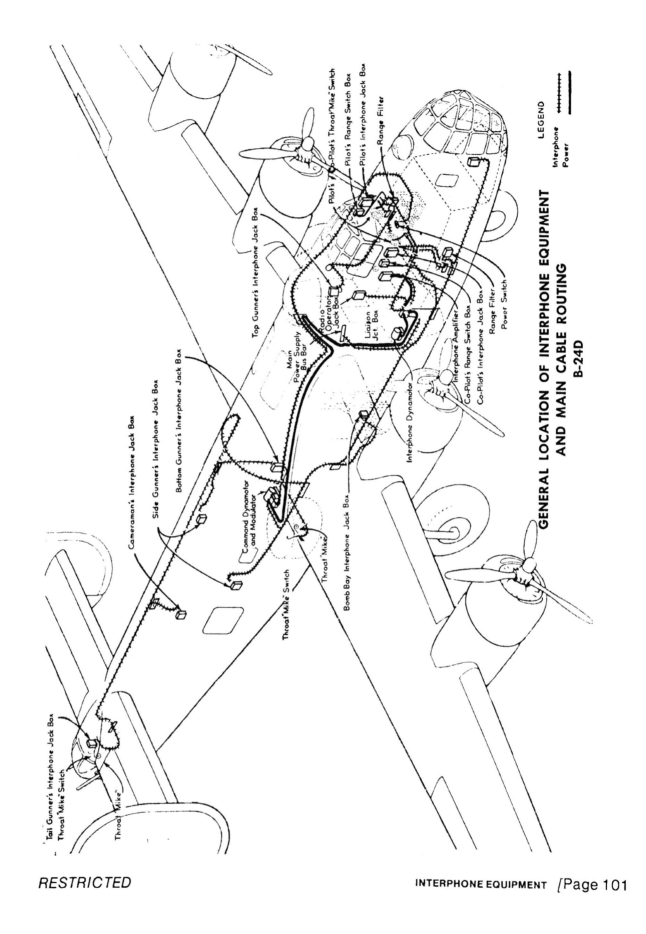

GENERAL LOCATION OF INTERPHONE EQUIPMENT AND MAIN CABLE ROUTING B-24D

INTERPHONE EQUIPMENT / Page 101

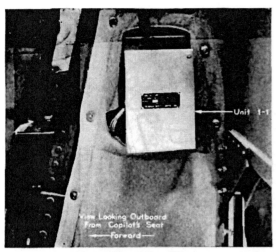

Interphone Amplifier

Since no provision is made to take power for the interphone system from the command equipment, a separate supply is provided and a separate amplifier is used. The amplifier, Unit 1-1, known as Type BC-347, is installed on the side of the Pilot's Compartment, directly aft of the Co-Pilot's seat. (See photo at left.) The dynamotor, Type PE-86, used for interphone power supply, is mounted on the floor alongside the Radio Operator's Table. The interphone system is turned "ON" or "OFF" by switch (3) of Unit 2-8, Page 111, simultaneously with the Command Transmitter.

Interphone System Operation—Each interphone jack box has five positions to which the selector switch may be adjusted along with a manual volume control. From these five selector positions the following may be accomplished: (See photo below, for interphone jack box markings.)

Position 1—"Compass." The audio output of the "Compass Receiver" only will be heard. A limited control of headset volume can be had by manipulation of the volume control. The microphone circuit is inoperative. This position is available at all stations.

Position 2—"Liaison." The liaison receiver output and the side tone of the liaison transmitter will be heard. A limited control of headset volume is possible with operation of the volume control. The microphone, "push-to-talk" switch operates the transmit-receive relay located within the liaison transmitter. The microphone will modulate the liaison transmitter when the microphone switch is closed, and the transmitter is in the "Voice" position.

Voice transmission from the "Liaison" position is available only from the Pilot's, Co-Pilot's and Radio Operator's Interphone Stations.

Position 3—"Command." The command receiver output and the sidetone from the command transmitter will be heard. A limited control of headset volume can be had by varying the volume control. The microphone "push-to-talk" switch operates the command send-receive relays which are located in the command receiver rack. The microphone will modulate the command transmitter when the "push-to-talk" switch is closed and the transmitter is in the "Voice" position. **This position is available at ALL inter-phone stations.**

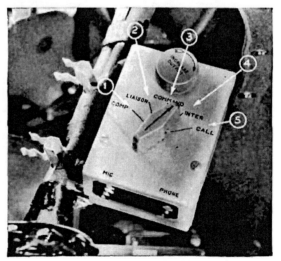

Interphone Jack Box

Flight Deck Controls

Unit 2-8, Command Transmitter and Interphone Power Control Box; Unit 4-3, Radio Compass Indicator. Unit 5-4, Marker Beacon Indicator; Units 7-1 and 7-2, Pilot's and Co-Pilot's Push Button Microphone Switches

Co-Pilot's Controls
Unit 1-4, Interphone Control; Unit 6-2, Filter Switch Box; Unit 7-2, Push Button Microphone SW; Unit 7-4, Microphone

Pilot's Controls

Position 4—"Inter." Provides an intercommunication system for use between crew members. The microphone connects to the input of the inter-phone amplifier, and the headphones to the output of this amplifier. **The volume control is not effective in this position.** This position is available at all interphone stations.

Co-Pilot's Filter

Position 5—"Call." This is an emergency call position in which **ALL** headphones at **ALL** boxes are in parallel across the output of the interphone amplifier. If an emergency should arise a crew member may contact any interphone station, even though it might be in use, by switching his jack box to the "Call" position. The microphone of the calling station is connected to the input of the interphone amplifier but leaves all other microphones in their respective positions.

Filter System—The RC-32 filter system includes two BC-345 switch boxes, Page 104. Units 6-1, 6-2 and two FL-5C filters Units 6-3, 6-4, as shown on this page. The filter is used for separating the voice (giving weather reports, etc.) from the beacon signal. The BC-345 switch permits the selection of beacon signal only, weather reports only, or beacon signal plus weather reports. This action is accomplished by switching to the desired marking: "Range," "Voice," or "Both." These devices are furnished for only Pilot and Co-Pilot.

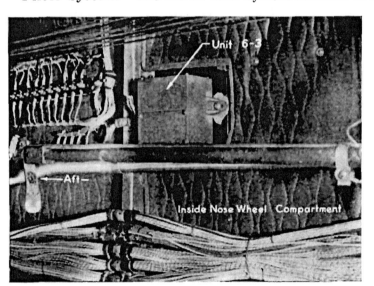

Pilot's Filter

RESTRICTED   INTERPHONE EQUIPMENT [Page 105

INTERPHONE STATION CONTROL

	Nose	Flight Deck				B. Bay	Tunnel				
	Bomber, Nav.	Pilot	Co-Pilot	Radio	Top Gun.	Bomb Bay	Bot. Tur.	Side Gun.	Side Gun.	Camera	Tail
Command											
Reception	X	X	X	X	X	X	X	X	X	X	X
Transmission	X	X	X	X	X	X	X	X	X	X	X
Headset Volume Cont.	X	X	X	X	X	X	X	X	X	X	X
Compass											
Reception	X	X	X	X	X	X	X	X	X	X	X
Transmission											
Headset Volume Cont.	X	X	X	X	X	X	X	X	X	X	X
Liaison											
Reception	X	X	X	X	X	X	X	X	X	X	X
Transmission		X	X	X							
Headset Volume Cont.	X	X	X	X	X	X	X	X	X	X	X
Interphone											
Reception	X	X	X	X	X	X	X	X	X	X	X
Transmission	X	X	X	X	X	X	X	X	X	X	X
Headset Volume Cont.											
Call											
Reception											
Transmission	X	X	X	X	X	X	X	X	X	X	X
Headset Volume Cont.											

INTERPHONE EQUIPMENT

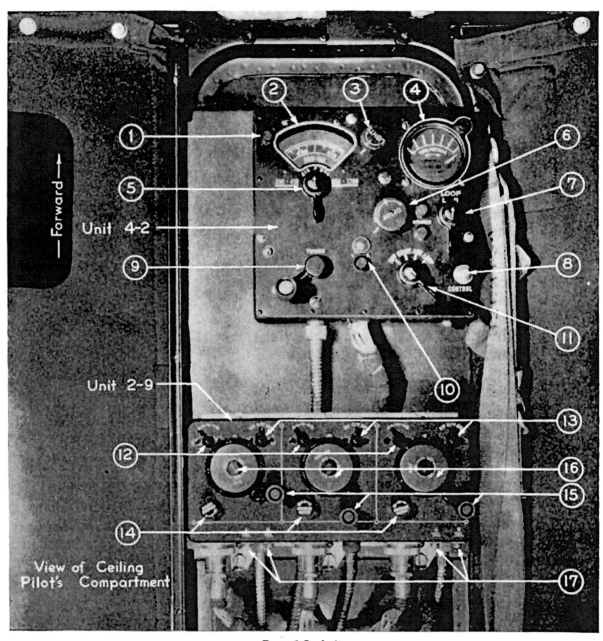

Top of Cockpit
Unit 2-9, Command Receiver Control Box; Unit 4-2, Compass Receiver Control Box

COMMAND EQUIPMENT

The "Command" equipment Unit 2, used in B-24D Airplane consists of the following units:

Unit 2-1 One BC-458-A Transmitter, Page 112
  2-2 One BC-459-A Transmitter, Page 112
      One FT-226-A Rack for 458 and 9-A Transmitters
      One FT-227-A Mounting for FT-226-A Rack
  2-3 One BC-453-A Receiver, Page 112
  2-4 One BC-454-A Receiver, Page 112
  2-5 One BC-455-A Receiver, Page 112
      One FT-220-A Rack for BC-453-4- and 5-A Receivers
      One FT-221-A Mounting for FT-220-A Rack
  2-6 One BC-456-A Modulator Unit, Page 110
      One FT-225-A Mounting for Modulator Unit
  2-7 One BC-442-A Antenna Switching Relay, Page 112
      One FT-229-A Mounting for Relay
  2-8 One BC-451-A Transmitter Control Box, Page 103, 111
      One FT-228-A Mounting for Control Box
  2-9 One BC-450-A Receiver Control Unit, Page 107
  2-10 One DM-33-A Dynamotor, Page 110
  2-11 Q6202-16 Terminal Strip
  2-12 DM-32-A Receiver Dynamotors

Necessary cordages and flexible control shafts for proper inter-connection and operation of the equipment.

Since both mechanical and electrical control of the Command equipment is normally delegated to the Pilot and Co-Pilot, the receiver control unit, BC-450-A (Unit 2-9) is mounted on the top of the cockpit between the Pilot and Co-Pilot; and the transmitter control unit, BC-451-A (Unit 2-8) is mounted on the pedestal.

The command set is a short range communication system used primarily for ship to ship communication. The BC-458-A transmitter has a frequency range of 5300 to 7000 KC, while the BC-459-A has a frequency range of 7000 to 9100 KC. The three receivers, namely BC-453-A, BC-454-A, and BC-455-A cover frequency ranges of 190 to 550, 3000 to 6000, and 6000 to 9100 KC respectively. No spare coils are needed for either transmitters or receivers.

The BC-453-A, BC-454-A, and 5-A receivers Units 2-3, 2-4, 2-5, and the BC-458-A and BC-459-A transmitters Units 2-1, 2-2 **(Page 112)** are mounted above the wing center section

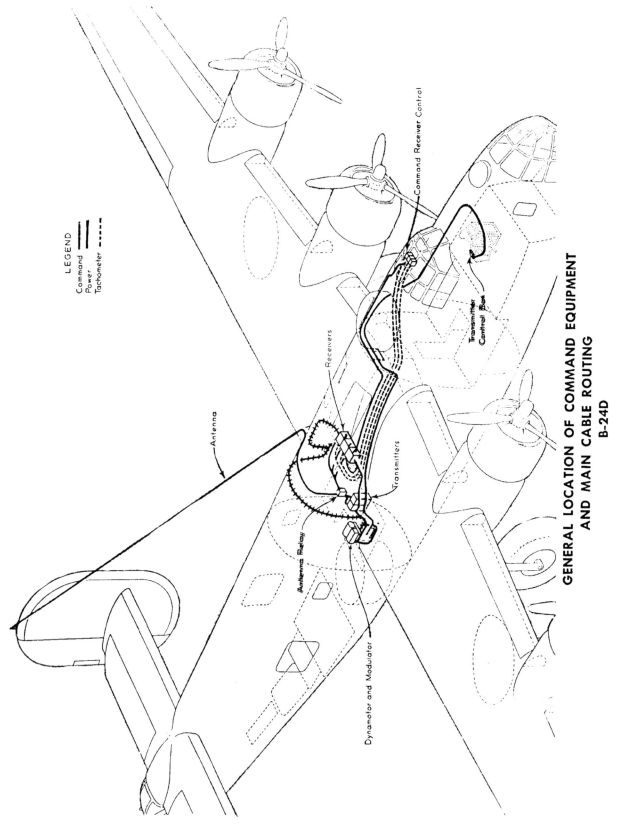

GENERAL LOCATION OF COMMAND EQUIPMENT AND MAIN CABLE ROUTING
B-24D

just aft of the life raft area. The BC-456-A modulator unit (Unit 2-6) and the DM 33-A Dynamotor (Unit 2-10) are mounted aft of the compass receiver on the rack for the RADAR equipment (See photo below).

A terminal strip, Unit 2-11, connects the transmitter remote control unit with the modulator unit, transmitter side tone, receiver output and interphone system. It is located in the compartment above the wing center section, outboard of the Compass Receiver unit and contains terminals only.

The receiver control box (Page 107) Unit 2-9 is divided into three identical control sections, except for dial calibrations, each connected to its own receiver. Thus, command receivers can be used individually or in any combination desired by the operator. A switch (13) is located in the upper right corner of each control section and has "CW," "MCW," and "OFF" positions. Two phone jacks marked "A-Tel," and "B-Tel" (17) are located on the aft side of the receiver control box through which receiver output and command transmitter side tone may be heard. These jacks are not normally used since the Command set is connected to the interphone system, and transmission or reception may be accomplished from any interphone station throughout the airplane as long as the Command equipment is turned "ON." Output to these jacks is controlled by the switch (12) in the upper left corner of each control marked "A"—"B." For interphone station operation turn all three switches to position "A." On the face of the BC-451-A transmitter control Unit 2-8 (Pages 103, 111), which is mounted on the pedestal, are three switches. The center switch (3) is marked "Trans. Power," "ON," "OFF." The switch (1) on the left has three positions marked

Command Dynamotor and Modulator

Command Transmitter Control Box

"Tone-CW-Voice." On "Tone" a keyed signal modulated at approximately 1000 cycles per second can be transmitted. "CW" is for keyed transmission of an unmodulated signal. The third position is for voice transmission. Two positions on the right hand selector switch (2) connect to the two command transmitters. The other two positions are not used.

On both "CW" and "Tone" positions, the microphone is inoperative but the push-to-talk switch may be used for keying the signal, or the key (6) on top of the BC-451-A control may be used. In addition to the two methods so far mentioned, an external or separate key may be plugged into the jack marked "Key" (5) on the aft side of the control box. Next to the key jack is another jack marked "Mic" (4) for plugging in a microphone on the Command set only.

Liaison Transmitter

Page 112] COMMAND EQUIPMENT

Each transmitter is supplied with a special frequency checking circuit and a plug-in crystal resonator. This crystal and its circuit are used for checking the frequency at a definite point on the dial only. **The crystal does NOT control the frequency being transmitted.**

Operation of the Command Set is not complicated, but a complete study of the Instruction Book, on the SCR-274-N Radio Set, should be made before operation or service is attempted.

Liaison Receiver

LIAISON RADIO EQUIPMENT

Unit 3—The Liaison Radio Equipment is identified as SCR-287-A equipment and includes the following units:

United 3-1—One BC-375-D Transmitter. Page 111.

    3-3 One TU-26 Transmitter Tuning Unit Frequency Range 200 to 500 KC
    3-4 One TU-5B Transmitter Tuning Unit Frequency Range 1500 to 3000 KC
    3-5 One TU-6B Transmitter Tuning Unit Frequency Range 3000 to 4500 KC   See
    3-6 One TU-7B Transmitter Tuning Unit Frequency Range 4500 to 6200 KC   Page
    3-7 One TU-8B Transmitter Tuning Unit Frequency Range 6200 to 7700 KC   119
    3-8 One TU-9B Transmitter Tuning Unit Frequency Range 7700 to 10000 KC
    3-9 One TU-10B Transmitter Tuning Unit Frequency Range 10000 to 12500 KC

    NOTE: Units 3-3 to 3-9 inclusive are removable units installed in Unit 3-1 to change frequency range.

    Six CS-48 Stowage cases for transmitter tuning units.

3-11 One BC-306-A Antenna tuning unit. Page 116.
3-12 SCR-211-D Frequency meter. Page 116.
3-2 One BC-348-H Receiver. Tuning range 1500 KC to 18,000 KC. Page 113.
3-10 One PE-73-C Dynamotor (Liaison). Page 118.
3-13 One J-37 Transmitting Key. Page 113.
3-14 One RL-42 Antenna reel.
3-15 One MC-163 Antenna Fairlead.
3-16 One F-10 Trailing Antenna
    One W-T-7A, SC-D-3338 Antenna weight.
3-17 One X41-B10A16 Antenna Transfer Switch. Page 116.
3-18 One BC-461 Antenna reel control box. Page 113.

These units are located through the airplane as follows:

BC-375-D Transmitter Unit 3-1—Under radio table, right side of flight deck behind Co-Pilot.
Two CS-48 Stowage boxes for transmitter tuning units—Left rear of flight deck.
Two TU Tuning Units in flight deck stowage.
Four CS-48 Stowage boxes—Above bomb bay, left side, aft of wing center section.
Four TU units in stowage above bomb bay.
One BC-348-H Receiver—On operator's table behind Co-Pilot.
One PE-73-C Dynamotor—Under flight deck, forward of anti-icer motors.
One J-37 Key—On operating table.
One RL-42 Antenna reel—Under flight deck, forward of dynamotor on flight deck floor brace.

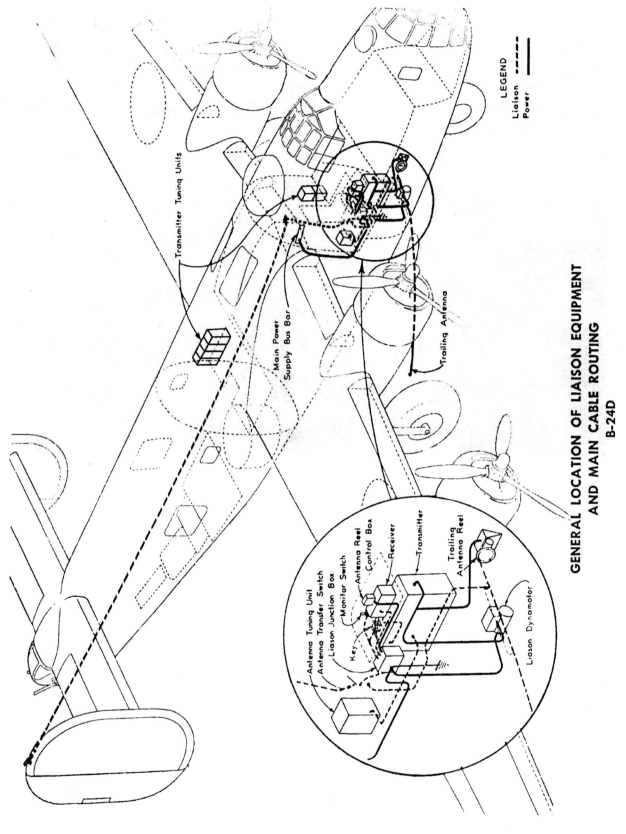

GENERAL LOCATION OF LIAISON EQUIPMENT AND MAIN CABLE ROUTING
B-24D

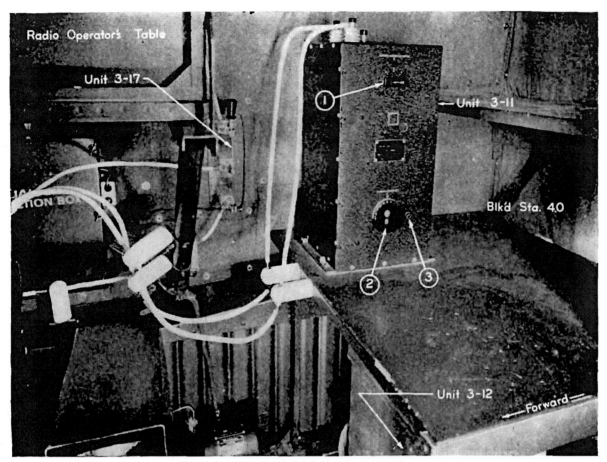

Right Side Flight Deck, Aft of Radio Table

One MC-163 and F10 Antenna Fairlead and trailing antenna—Under flight deck, right side.

One BC-306-A Antenna Tuning Unit—Right rear of flight deck, aft of radio table.

One SCR-211-D Frequency Meter—Right rear section of flight deck. Portable equipment used for frequency checking.

BC-461 Antenna reel control box—Right wall of cabin above radio table.

X41-B10A16 Antenna Switch—Right wall of cabin, aft of radio table.

The liaison equipment is used generally for long distance communication between the airplane and base or airplane and ground stations. It is used primarily for reporting ship position or flight progress. Control of liaison equipment is held by the Radio Operator.

Liaison Equipment Operation—The liaison receiver face contains nine controls, as follows (see Page 128):

1. C.W. Oscillator—"OFF" and "ON" Switch—upper left.
2. Crystal "IN" and "OUT" Switch—to the right of C.W. Oscillator switch.
3. Power switch: ("A.V.C.") or ("M.V.C.") below oscillator switch.
4. Volume Control—lower left side of face.
5. Beat frequency pitch control—to the right of volume control.
6. Band change switch—lower portion of dial.
7. Tuning control—below and to the right of dial assembly.

8. Dial light brilliancy control—upper right.
9. Antenna Alignment—extreme right center.

In addition to the above nine controls, there are four connectors located on the face of the receiver. At the extreme right in the lower corner are the terminals for the antenna and ground (10). At the extreme lower left are two jacks for headset attachment plugs PL-55 (11).

The liaison receiver is in operation when the switch (3) is in either "AVC" or "MVC" position. The tuning control is operated with the power switch in the "MVC" position. After location of the desired signal, adjust volume to desired intensity and throw power switch to

Liaison Transmitter, Under Radio Table

RESTRICTED LIAISON RADIO EQUIPMENT [Page 117

"AVC" for automatic maintenance of desired volume. **For reception with the liaison receiver, the "Monitor Switch"** (12) mounted on the operator's table between the J-37 key (3-13) and the receiver, **must be in the "Normal" position.** The "Monitor" position is used when it is desired to operate the liaison receiver for a check on the liaison transmitter tuning or adjustment. For further details on this check, see "Liaison Transmitter" which follows.

Band switching is accomplished by turning of the control knob (6). The band in use is indicated through the dial mask.

The receiver dynamotor is contained within the receiver and is only accessible by removing the receiver from its case.

Liaison receiver output is available at any interphone junction box in the airplane.

For CW reception the beat frequency oscillator with its manual pitch control may be used either with or without the crystal filter at the option of the operator.

LIAISON TRANSMITTER—The liaison transmitter Unit 3-1, **Page 117,** has the following controls and indicators on the front of the panel:

1. Power Switch
2. Filament Voltmeter
3. Voltmeter Switch
4. Master Oscillator Tuning Controls
5. Test Key
6. Output Control Switch
7. Plate Current Meter
8. Antenna Current Meter
9. Antenna Tuning Control (Inductance)
10. Antenna Circuit Switch
11. Power Amp Tuning Dial
12. Antenna Tuning Control (Capacity)
13. Band Change Switch
14. Antenna Coupling Switch
15. Antenna Inductance Selector Switch

Liaison Dynamotor

The Liaison Dynamotor Unit 3-10 (see photo above) is mounted under the flight deck right side, and supplies high DC plate voltage to the transmitter.

The Liaison transmitter Unit 3-1 **Page 117,** is turned "ON" and "OFF" by the power switch (1) mounted on the face of the transmitter. **When the transmitter is "ON" it is important that the filament voltage be closely maintained at 10 volts** as indicated on the voltmeter (2). The C.W. and Modulator filaments are checked individually by means of a selector switch (3).

Each tuning unit, **Page 119** contains the necessary tuned or tunable circuits to permit operation on a frequency band within the limits as specified on the calibration chart attached to the front of the tuning unit. The settings as shown on this chart are very close with regard to accuracy but with the Monitor switch (12), **Page 113,** thrown to the "Monitor" position the side tone of the transmitter is cut off and the output frequency can be tuned to exactly match (zero beat) with the desired frequency setting on the liaison receiver

Left Rear Flight Deck

Additional Spare Tuning Units—Top Deck

RESTRICTED   LIAISON RADIO EQUIPMENT /Page 119

(Unit 3-2). During this tuning, the receiver should be "ON" and the C.W. switch (1) also thrown to the "ON" position. With both transmitter and receiver "ON" in the above positions and the monitor switch (12) **Page 113**, in the "Monitor" position, press the transmitter key Unit 3-13 and tune the transmitter oscillator frequency dial (4) until the transmitter output is heard in the receiver. Trim the transmitter adjustments for maximum output and recheck on the receiver. Transmitter is now adjusted for operation and standing by, ready for break-in operation on the station or frequency to which it has been adjusted.

CAUTION: DO NOT MAKE AN INTERNAL ADJUSTMENT OR REPLACEMENTS WITH TRANSMITTER DYNAMOTOR RUNNING.

In order to tune the antenna for transmission on the lower frequencies (below 800 KC a BC-306-A antenna tuning unit, **Page 116,** has been provided and installed at the rear of the flight deck right side, with its connections close to those of the liaison transmitter. Since a variation of connections is necessary for best operation follow chart below for proper operation.

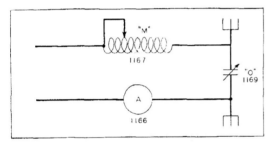

View showing internal connections for No. 1 position.

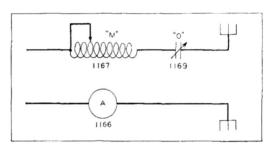

View showing internal connections for No. 2 position.

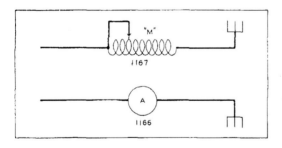

View showing internal connections for No. 3 position.

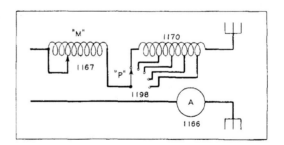

View showing internal connections for No. 4 position.

ANTENNA INDUCTANCE SELECTOR SWITCHES
B-24D

CRUISE CONTROL CHARTS
and Tactical Cruising Problems

When an airplane leaves on a mission it has definite objectives. To accomplish these objectives it must:

1. Carry a certain load.
2. Carry a sufficient supply of gasoline.
3. Travel at a certain average speed and travel a certain distance.

The cruising control chart dispenses with the use of complicated formulae, slide rules, tables, etc., in the solution of problems involving: air density, airspeed indicator corrections, and weight variations of speed and power for the B-24D Airplane. A brief study of the components making up the chart should assist the Pilot in its use.

The facts on which it is based are as follows:

1. Changing air density (P) changes the amount of lift and drag produced. Air density depends on pressure and temperature. **Therefore, we must know the density altitude to obtain the power required.**
2. The airspeed indicator reading is affected by air density. It does not read true airspeed except at standard sea level air density. **So, a conversion must be made to obtain true airspeed from the indicated airspeed.**
3. Increasing weight increases the amount of lift required for a given set of conditions. Lift coefficient must be increased, and as a result drag coefficient increases. **Increasing weight increases the power required to fly at a given speed, or decreases the speed obtained with a given amount of power.**

Taking the individual parts of the cruising control chart in the order listed, we may see how each contributes to the final answer.

1. **Density Altitude:** First, take the case of air density. At standard temperature the pressure altitude and density altitude are the same. If the temperature is above standard, the density altitude is greater than the pressure altitude. If the temperature is below standard the density altitude is less than the pressure altitude. Examples may help to clarify this.

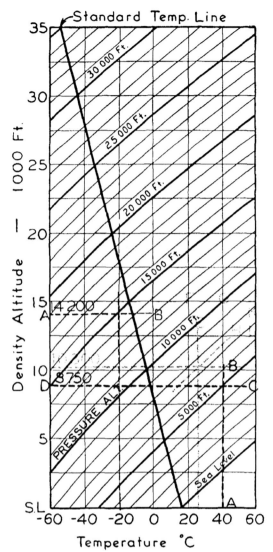

Pressure and Density Altitude

Example 1:
Pressure altitude = 5000'
Temperature = 40° C.
Read from chart—density altitude = 8900'

To use chart: Select temperature line 40°—vertical line AB. Follow this line until it intersects the **pressure** altitude 5000 foot line CD—(interpolate for intermediate altitude where necessary). (See Example 3 below.) Extend this intersection point to the left to the density altitude line and read; density altitude 8900'.

Example 2:
Pressure altitude = 15,000'
Temperature = −20° C.
Read from chart—density altitude "AB." = 14,400'

Example 3:
Pressure altitude = 7500'
Temperature = 25° C.
Interpolate line E between F and G
Interpolate line H between K and L
Extend intersection of E and H to density altitude and read density altitude = + 10,300'

NOTE: Pressure altitude is the altimeter reading when the barometric scale on the altimeter is set to 29.92" Hg. (1013 millibars).

AIRSPEED ALTITUDE CORRECTION

NOTE: The location of the pitot static heads on the B-24D Airplane is such that there is a small position error in indicated airspeed which is the same for all airplanes of this model regardless of the instrument error. A calibration of the Pilot's Airspeed Indicator for each airplane should be made in an instrument laboratory. The position error has been taken into account in all the charts in this manual so that only the correction for the individual Pilot's instrument error need be made.

Conversion from indicated airspeed to true airspeed for any density altitude may be accomplished by using this chart.

Example 1: Pilot's indicated airspeed = 160 MPH
Density altitude = 7500'
Find true airspeed = 175 MPH
On chart, follow 160 MPH IAS line AB to intersection E with 7500' density altitude line CD. Read true airspeed 175 MPH on curve EF.

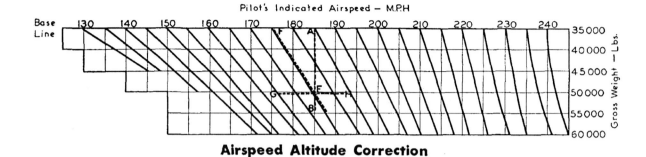

Airspeed Altitude Correction

WEIGHT—SPEED VARIATION CHART

2. **Weight Variation:** If we know how fast a plane will go with a given HP at one gross weight, the gross weight correction lines on the chart shown here will tell us how fast the same plane will go with the same HP if the gross weight is changed. Remember: Increasing weight decreases speed with the same power, or if constant speed is desired, the higher weights require higher powers.

Example 1:
    Indicated airspeed 185 MPH at 35,000 lbs. gross weight
    Find indicated airspeed for 50,000 lbs. gross weight.
    Drop line AB to 50,000 lb. line GH
    Extend EF from intersection to scale and read 176 MPH.

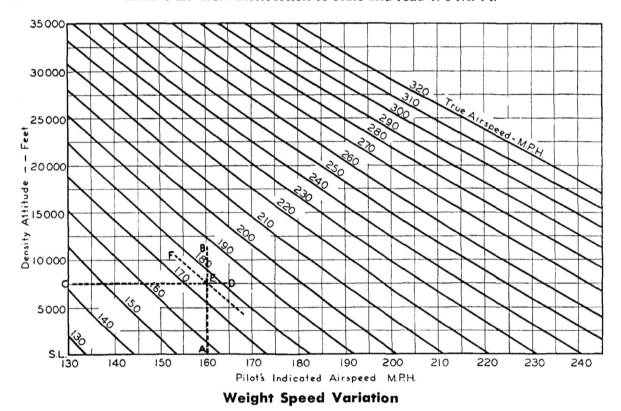

Weight Speed Variation

Now, combining these three charts as shown on the Composite Cruising Correction Chart on **Page 124**, one may see the basis on which the cruising control chart is established.

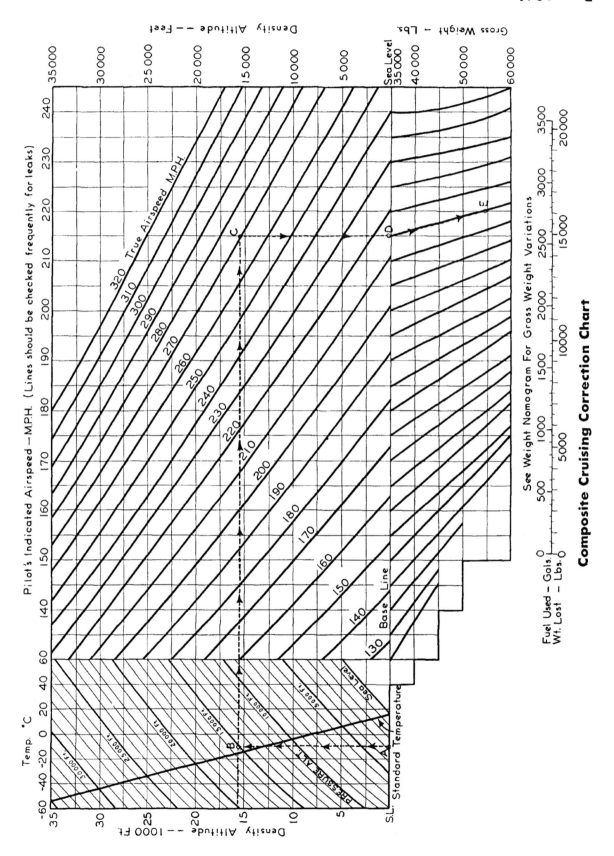

Composite Cruising Correction Chart

THE COMPLETE CRUISE CONTROL CHART

(Document 32-1-3)

On the final chart lines have been added to the airspeed chart which show the BHP requirements for the basic gross weight. This, with the other two correction charts, makes the composite cruising control chart. Its use allows quick and direct solution of problems involving various combinations of: altitude, temperature, airspeed, gross weight, engine RPM, manifold pressure and fuel consumption. It also gives approximate speeds for the maximum range operation.

> NOTE: The curves which give density altitude involve only the basic laws of the atmosphere and are applicable to any airplane.

The Pilot's airspeed indicator reading requires correction only for the instrument error of each individual airplane. The position error, which is constant for all airplanes of this model, has been taken into consideration in the Cruising Control Chart.

To Determine BHP Required for Any Desired True Airspeed at Any Gross Weight:

Enter chart with density altitude (determined from pressure altitude and temperature). Project horizontally to true airspeed for which flight is charted. Project point vertically downward to the 35,000 lb. gross weight (base line). Follow line parallel to gross weight calibration lines to intersection with line of airplane gross weight. Project intersection point vertically to charted altitude and read BHP, RPM, manifold pressure and fuel flow by interpolation.

Example:

Given Pressure Altitude	= 15,000′
Temperature	= −10° C
True airspeed 270 MPH (required to make good a previously calculated ground speed).	
Gross weight	= 55,000 lbs.

Solution:

Enter the Density Altitude Temperature Chart with 15,000′ and −10° C. and find the Density Altitude = (AB) 15,600′.

Extend this line to the true airspeed of 270 MPH. Read the Pilot's indicated airspeed = (BC) 215 MPH. To this apply the instrument correction to obtain the airspeed which the airspeed meter will indicate.

Drop a vertical (CD) to the 35,000 pound line and follow the gross weight speed change (DE) curve to the 55,000 pound line.

Extend this vertically (EF) to the density altitude 15,600 foot line and read power required = 97% rated power. The following are engine instrument readings; RPM 2520; manifold pressure 44.5; fuel consumption 570 GPH.

WARNING: The airspeed for a given power and weight is reduced by opening cowl flaps. (Approximately 0.8 MPH reduction in indicated airspeed for each degree of cowl flap opening in an average cruising condition). Keep constantly aware of weather conditions. A turbulent front is extremely hazardous to a heavy, fast airplane. Gusts encountered under these conditions may impose stresses greater than those for which the airplane structure was designed.

FOR USE IN CRUISING FLIGHT

Determine density altitude and obtain desired true airspeed in normal manner using free air temperature, pressure altitude and airspeed indicator calibration correction. Set manifold pressure and RPM to charted values as required to give speed desired. At charted MP and RPM, TAS will be low in hot weather, high in cold, when compared to charted values. Jockey power slightly as required (increase MP to increase speed, decrease RPM to decrease speed) until charted TAS is obtained. This will establish power exactly. Fuel flow will thereby be established. Do not increase manifold pressure more than 2 in. above charted values without raising RPM. For automatic rich operation do not exceed 35.5 in. manifold pressure nor 2325 RPM for continuous cruising. For automatic lean operation **do not exceed** 32 in. manifold pressure nor 2200 RPM for continuous cruising.

For steady cruising, it should not be necessary to reset power more often than each hour, every three hours will probably be satisfactory.

A scale is given at the bottom of the cruising control chart for approximation of gross weight changes due to fuel consumption. For more accurate determination of gross weight after use of expendable items of weight, see the weight nomograms.

ENGINE LIMITATIONS FOR R-1830-43 ENGINES:

Conditions	RPM	LIMITS Manifold Pressure Inches Hg	Time
Take-off	2700	48.5	5 Min.
Military Power	2700	48.5	5 Min.
Normal Rated Power	2550	46	1 Hour
Max. Cruising Auto Rich	2325	35.5	Continuous
Max. Cruising Auto Lean	2200	32	Continuous

Note I: A tight exhaust system is necessary for good turbo operation. Poor spark plugs reduce power—require excessive manifold pressure.

Note II: Decrease manifold pressure 1.5 in. per 1,000' increase in altitude above 25,000'.

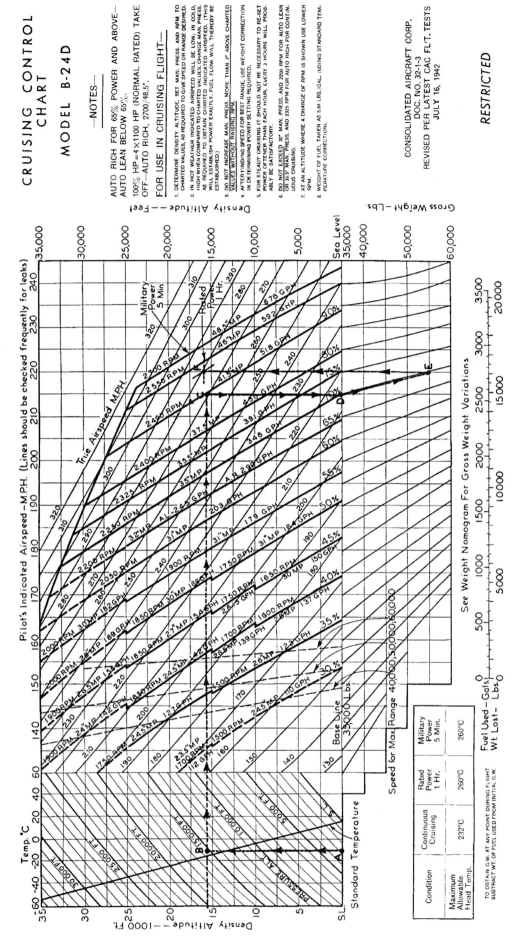

INSTRUCTIONS FOR USE OF
MAXIMUM RANGE CONTROL CHART
(Document 32-2-4)

Enter the chart with the desired gross weight, using the scale at the bottom of the chart. Project vertically, and at the proper altitude for each set of curves, read in turn:—

1. Airspeed (to be corrected for instrument error)
2. Engine RPM
3. Manifold Pressure
4. Total fuel consumption
5. Miles per gallon

Having picked off the conditions, the RPM is set first, then MP is varied to give the desired airspeed. At charted speed and RPM, the MP will be high in hot weather, low in cold weather, when compared to charted values. MP should not be raised more than 2 ins. above the charted value without raising RPM.

Example: (Taken from Maximum Range Control Chart)

Given: Gross Weight—45,000 lbs.; Density Altitude—15,000 ft.

Results: Pilot's Indicated airspeed—147 (apply Pilot's Instrument correction)
RPM = 1690 Manifold Pressure = 26.5 in. Hg. (approx.)
BHP = 440 per engine (approx.) Fuel flow = 137 GPH (approx.)
Miles per gal. = 1.32 (approx.)

Note I: For steady cruising it should not be necessary to reset power more often than each hour. Every three hours will probably be satisfactory.

Note II: At indicated airspeeds less than 140 MPH, when flying on the A.F.C., the Pilot should pay close attention to the airplane in order to prevent inadvertent stalling when the airplane flies through sharp updrafts. However, in cases where maximum range and endurance demand low speeds, the airplane may be flown manually; returning to Automatic Control when the low speeds are no longer required.

Note III: At speeds other than those for maximum range or maximum endurance, the cruising control chart is used as a guide to obtain power, RPM, and manifold pressure.

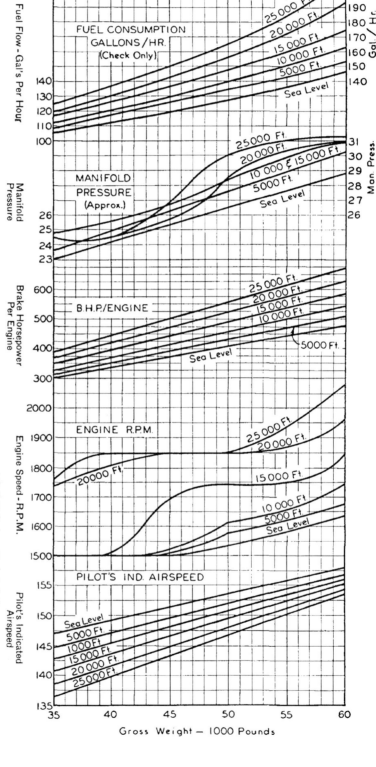

THE TAKE-OFF CHART
(Document 32-3-1)

NOTE: The take-off run given by this chart includes distance to clear a 50 foot obstacle. Ground run may be obtained as noted on chart.

Correction to density altitude for this chart is made in the same manner as on the cruising control chart. The main body of this chart gives take-off run for the case of a hard-surfaced runway and no wind.

The take-off tests were carried out in the following manner: After initial warm-up period each engine was run up separately to 2700 RPM and 49 ins. Hg manifold pressure (take-off settings). The throttle was then pulled back, leaving RPM and turbo controls set. When ready for the take-off, the brakes were held as the throttles were opened. As soon as the tachometers showed 2100 RPM, the brakes were released, allowing the ship to start forward. Throttles were then pushed all the way forward during acceleration. This method puts less strain on the plane than when the brakes are held until full power is reached, and gives little if any increase in the take-off run.

For the example illustrated on the chart:

(Temperature = 10° C., pressure altitude 5000 ft. and gross weight = 45,500 lbs.). The take-off run over a 50 foot obstacle on a hard-surfaced runway with no wind would be 2750 ft. reading on the scale at the top or bottom of the chart.

To make correction for different ground surface condition the point of intersection of density altitude and gross weight is projected vertically to the base of the main chart and then parallel to the diagonal lines to the surface condition to be encountered. With no wind, this point would give the take-off run. Using the above data and soft ground, the take-off run is found to be 4060 ft.

Wind correction is made by using the lowest set of diagonal lines after projecting the point last found vertically to the O wind line. Following parallel to these lines to 20 MPH wind velocity for the above data, find the take-off run (= 3100 ft.).

If the take-off had been from a hard-surfaced runway into a 20 MPH wind, the run would have been 2000 ft.

If for any reason a downwind take-off is contemplated, the approximate distance can be calculated by projecting vertically from the ground condition point to the horizontal line of wind velocity, then following parallel to the diagonal lines (upward and to the right) to the O wind line and project this point to the base scale, e.g., for the case illustrated, take-off with a 10 MPH tail wind would require 4630 ft. (1530 ft. longer run than with the 20 MPH head wind), and from the hard surfaced runway a 10 MPH tail wind would require 3200 ft., or more than a 50% increase over the distance with 20 MPH head wind.

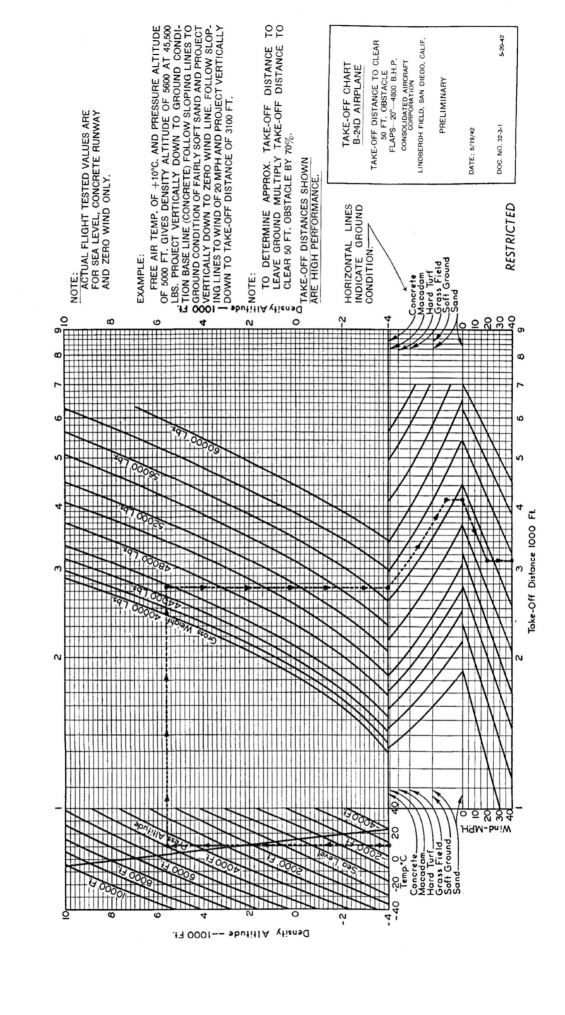

AMERICAN-BRITISH TERMINOLOGY

AMERICAN	BRITISH
Battery	Accumulator
Control Stick	Control Column
Cotter Pin	Split Pin
Firewall	Fireproof Bulkhead
Vertical Stabilizer	Fin
Surface Controls	Flying Controls
Surface Control Lock	Locking Gear
Wing	Main Plane
Clevis Pins and Clevises	Pins and Shackles
Mooring Rings	Picketing Rings
Left	Port
Pitot Tube	Pressure Head
Right	Starboard
Horizontal Stabilizer	Tail Plane
Empennage	Tail Unit
Landing Gear	Undercarriage

AMERICAN-BRITISH METRIC UNITS

The following general table of conversions may be used where calculations are necessary:

Multiply	By	To Obtain
U. S. Gallons (gal.)	0.833	(Imp. gal.) Imperial Gallons
U. S. Gallons	3.785	(l) Liters
Miles per hour (mph)	1.609	(KmPH) Kilometers per hour
Miles per hour	0.8684	Knots
Miles	1.609	(Km) Kilometers
Miles	0.8684	Nautical Miles
Feet (ft.)	0.3048	(M) Meters
Inches (in.)	2.54	(cm) Centimeters
Pounds (lb.)	0.4536	(Kg) Kilograms
Pounds per sq. in. (lbs./sq. in.)	0.0703	(Kg/sq. cm) Kilograms per square Centimeter
Inches of Mercury (in. Hg.)	2.54	(cm Hg.) Centimeters of Mercury
Horse Power (h.p.)	1.014	(m.h.p.) Metric Horse Power
Degrees Centigrade (°C) + 17.8	1.8	(°F) Degrees Fahrenheit

APPENDIX

HOW TO FLY THE B-24D AIRPLANE

RESTRICTED

HOW TO START – TAXI – WARM-UP – TAKE-OFF AND LAND THE B-24D AIRPLANE

THE B-24D AIRPLANE—The B-24 is a large airplane but it is neither difficult nor tricky to fly. The operation of any airplane, single or multi-engine, requires certain operations performed in a definite sequence. This sequence, as it applies to the B-24, is covered by a complete and explicit Pilot's Check-off List. This Check List must be followed exactly and intelligently. Even the most experienced Pilots can not remember a long detailed list. The Pilot's Check List, as included in the airplane and in this "Pilot's Flight Manual," is for the experienced B-24 Pilots as a reminder and a sequence check-off. Certain other ordinary routine operations, which the experienced Pilot does instinctively, must be performed in connection with this list and it is the purpose of this text to fill in these gaps so that the inexperienced Pilot who has never seen a B-24 can learn the entire operation in detail and follow the Check-off List intelligently. We will follow through each step from the time the crew first approaches the airplane until the airplane returns to the field from its flight.

OPENING THE BOMB BAY DOORS—As the crew approaches the airplane the Flight Engineer opens up by unlocking and opening a small access door on the right side of the fuselage. Reaching in through this door he opens the bomb doors by pulling outward on the handle of the auxiliary bomb door valve.

CHECKING REMOVAL OF THE PITOT COVERS—While this is being accomplished the seasoned and wise Pilot steps forward to make doubly certain that the covers on the the pitot heads have been removed. He knows that these covers can not be removed in the air and that with them "ON" the Airspeed Indicator is inoperative. An airplane of this size can not be operated safely by guessing airspeeds. With the ship open, the Pilot and Co-Pilot enter first and go forward to the cockpit.

TURN THE FUEL VALVES "ON"—The Flight Engineer then enters. His first act is to turn on the four fuel selector valves, one for each engine. These are labeled to connect an engine to a corresponding numbered system or to the cross-connection to which all tanks units and engines may be connected. These valves are located overhead, forward of the main center section spar, two on the right of the centerline controlling the flow to Engines 3 and 4; and two on the left side controlling the flow to Engines 1 and 2. They are so numbered, 1, 2, 3, and 4.

CHECK THE QUANTITY OF FUEL IN THE TANKS—The Engineer next checks the fuel load. This is shown by the two visual gauges located on the left forward face of the bomb bay bulkhead at the rear of the flight deck. Each gauge is connected by a two-way valve with two of the four main fuel systems so that by operation of these valves the quantity of fuel in each of the four systems may be determined; 1 and 2 on the outboard gauge, 3 and

4 on the inboard gauge. In taking a reading the Inclinometer, located outboard of the gauges, must read zero for accurate reading. A reading off zero can cause a high percentage of inaccuracy in the quantity of fuel aboard. The figures on the gauges read in U. S. Gallons. Each system should have a minimum of 300 gallons, giving a total for the four systems of 1200 U. S. Gallons.

THE PILOT AND CO-PILOT SEAT ADJUSTMENT—The Pilot and Co-Pilot in their seats adjust them for maximum comfort by adjusting levers located outboard of each seat. The three levers permit adjustment fore and aft, up and down, and tilt. With the seats adjusted properly for comfort the rudder pedal adjustment for proper length is the next step. Each pedal adjustment ratchet is located between the pedals. Adjustment is accomplished by pushing the ratchet lever away from the pedal with the toe and moving the pedal fore or aft to the proper position. Be sure the latch is engaged properly.

REMOVE THE CONTROLS LATCH AND CHECK THE CORRECT SURFACE MOVEMENT AND CONTROLS VISUALLY—The controls lock which holds the locking lever "UP" in the locked position is removed and the strap is stowed securely in the overhead. It is wise to check the locking lever in the **"FULL DOWN"** position to make sure it has dropped all the way and that the lock is released. With the controls lock released, check the movement and direction of the controls in the normal manner by turning the wheel for the ailerons; by exercising the wheel fore and aft for the elevators; and by pushing each rudder pedal for the rudders. As the Pilot turns the wheel the Flight Engineer or Co-Pilot should check the movement of the controls visually and call out to the Pilot the direction of movement. This check might seem superfluous but controls have been found crossed after re-rigging.

ENGINES MUST BE PULLED THROUGH BY HAND BEFORE STARTING— Before beginning the actual starting sequence, each engine must be pulled through by hand to check the free turning of the engine, and to clear any oil or fuel which may have accumulated in the combustion chambers, which if present, would most certainly result in a damaged engine.

BE SURE THE IGNITION SWITCHES ARE "OFF" BEFORE ENGINEER APPROACHES ENGINES—Before the Engineer approaches the engines, ignition switches must be definitely checked "OFF" and the master ignition switch must be checked "OFF." A kick-back would result in serious injury to the Engineer. The Engineer must pull each propeller through six blades which completes two full revolutions or one complete cycle. Even with the ignition switches checked in the "OFF" position, the Engineer should be constantly aware of the possibility of a broken ground wire which would cause a kick-back. Keep clear of the propeller plane of rotation while the propeller is being pulled through. Do not become careless. The ignition switch "OFF" can be just like the famous unloaded gun.

SOURCE OF ELECTRICAL POWER FOR TURNING THE ENGINES—Power for starting the engines can be drawn (1) from the ship itself, making use of the batteries plus the Auxiliary Power Unit; or (2) from outside the ship, making use of a battery cart or other outside power unit. When the ship's batteries are used, the Auxiliary Power Unit must be running unless it is inoperative. Without it, the direct strain of starting is a severe load on the ship's batteries and shortens their life. It may even do immediate damage to the plates of the batteries. If the Auxiliary Power Unit is inoperative, never attempt to use low batteries—they will not start the engines and will cause fusing of the starting solenoids as soon as they are closed. The Auxiliary Power Unit is never used alone. It does not have enough capacity to take care of the starter surge and the power requirements of the starters. The main difference between starting with ship's power and starting with outside power as far as operation is concerned is in the position of the battery switches located over the master switch on the right side of the cockpit. In case of a start with ship's power, these are turned "ON" connecting the ship's batteries and the Auxiliary Power Unit with the main power circuit. In case of a start with outside power, these battery switches are turned "OFF" since the outside battery plug connects directly to the main power circuit. Connection for the outside source of power is a receptacle reached through the same small door on the right hand side of the fuselage forward of the bomb doors through which the auxiliary bomb door handle is reached in first opening the airplane.

STARTING PROCEDURE—For starting either with ship's power or with outside power.

1. BATTERY SWITCHES IN PROPER POSITION—These switches located immediately over the master switch consist of three units in the earlier and two units in the later installations. All switches are turned "ON" when ship's power is used for starting and "OFF" when outside power is used.

2. GENERATOR SWITCHES "OFF"—The generator switches located on the generator panel on the forward face of the bomb bay bulkhead on the left side are turned "OFF" to prevent vibration damage to the voltage regulator.

3. AUXILIARY POWER UNIT—When the start is made on ship's power, the Auxiliary Power Unit is started. This unit is located on the left side under the flight deck. If ship's batteries are fully charged the unit may be started by priming and then pressing the starter button on the unit. The generator then acts as a starting motor to turn the gasoline engine of the unit. If the ship's batteries are low, the Auxiliary Power Unit must be started by hand and in this case the starting is accomplished before the ship's battery switches are turned "ON." Starting procedure for hand starting is to prime the unit, wind the starter rope around the drum and pull to turn the engine over.

TURN "ON" ALL FOUR IGNITION SWITCHES AND THE MASTER SWITCH—The IGNITION SWITCHES 4, 3, 2 and 1 and the Master Switch, located on the right side of the cockpit at the Co-Pilot's right hand are turned "ON."

AUXILIARY HYDRAULIC SWITCH "ON"—As the Engineer leaves the airplane to stand by the starting engines he reaches overhead on the right side ahead of the spar and turns "ON" the Electric Switch. The auxiliary pump furnishes hydraulic power until No. 3 Engine is started.

WHEEL CHOCKS—Wheel chocks are always used if available and are placed in front of each wheel.

PARKING BRAKES "ON"—The Pilot sets the parking brakes, even though the wheel chocks are installed, as a further precaution to keep the airplane from rolling when the engines are being started.

A.C. POWER FOR INSTRUMENTS IS TURNED "ON"—The Alternating Current or inverter power switch located on the left rear of the pedestal is turned "ON" to either No. 1 or No. 2 Inverters. These inverters furnish the Alternating Current power for the electrically operated engine instruments.

AUTOMATIC FLIGHT CONTROL—Automatic Flight Control must be "OFF" for take-off. If take-off is attempted with this unit connected, free operation of the controls is impossible. A slide bar on top of the Automatic Flight Control Unit on the left side of the pedestal is pulled aft to turn "OFF" all the Automatic Flight Control Switches.

SET ALTIMETER FOR THE FIELD BAROMETRIC PRESSURE—The Altimeter is set for the proper barometric reading, as obtained from flight operations control tower, by turning the adjustment below the face of the dial.

DE-ICER CONTROLS "OFF"—The De-Icer and Anti-Icer Controls to the right of the pedestal on the Instrument Panel are turned "OFF." De-Icer operating during take-off would change the entire wing lift characteristic and would be a hazard.

INTERCOOLERS MUST BE "OPEN" FOR STARTING—The intercooler shutter switches located forward on the pedestal in the center are placed in the "OPEN" position. Closed intercoolers would cause overheating and detonation.

COWL FLAPS "OPEN"—Cowl flaps are "OPEN" for starting to prevent excessive temperatures. Cowl flap switches are located on the right side of the pedestal. To operate place them in the "OPEN" position and hold until the cowl flaps are fully "OPEN."

PROPELLERS ARE SET TO HIGH RPM TO REDUCE CYLINDER PRESSURES AND SHORTEN WARM-UP—High RPM's are desirable to hasten warm-up and to reduce cylinder pressures. The propellers are set for High RPM by moving the switch bar of the switch gang located on the forward left side of the pedestal to the "INC" position and holding it so until the four indicator lights on the center of the Instrument Panel flash "ON."

SUPERCHARGERS MUST BE "OFF"—All four supercharger controls are set in the "OFF" position. This opens the engine exhaust tail gate. If the engines are started with turbos "ON" and the tail gate is closed the exhaust system would in all probability be blown off by the usual "pop" or exhaust explosion when the engines are starting.

MIXTURE CONTROLS MUST BE IN "IDLE CUT-OFF" BEFORE STARTING THE BOOSTER PUMPS—Mixture controls are placed in the "IDLE CUT-OFF" position. If they are in any other position when the booster pumps are started the engines will become flooded. This will create a fire hazard and make starting difficult.

AIRPLANE FIRE EXTINGUISHERS ARE IMMEDIATELY AVAILABLE IF NEEDED—The airplane fire extinguisher valves, located to the right of the Co-Pilot, should be checked for position so that they can be operated quickly if needed.

STAND BY THE STARTING ENGINE WITH A PORTABLE FIRE EXTINGUISHER—The Flight Engineer or ground mechanic always stands by the starting engine with a portable fire extinguisher. A flooded engine could result in a fire which could be extinguished immediately but which might otherwise be very serious.

NOW THE ENGINES CAN BE STARTED—Engines are started in the order: 3, 4, 2, and 1. They are started in this order, which is from inboard to outboard, so that the Engineer does not have to walk through or toward a moving propeller when standing by with the portable fire extinguisher. The No. 3 Engine is started first because it has the only hydraulic pump attached to it. The Co-Pilot, who usually starts the engines, checks visually to see that all personnel and obstructions are clear of the propellers, calls "ALL CLEAR" and is check answered "ALL CLEAR" by the Engineer.

TURN "ON" THE FUEL BOOSTER PUMPS FOR PRIMING PRESSURE—The four electric fuel booster pump switches, located under the Instrument Panel to left of Co-Pilot's Control Panel, are turned "ON." The booster pump pressure is required for priming the engines, in order to force fuel through the carburetor when the electric priming solenoids are "OPEN" for the priming switches, as the engine driven pumps are not operating until the engine starts. The booster pumps are further required on take-off, on landing, and at altitudes after the fuel pressure has dropped two pounds. Booster pumps insure a positive flow of fuel to the engine pump because they are located directly on the tank outlets.

ENERGIZE THE STARTER—While the No. 3 Engine is being primed with one hand, the Co-Pilot holds the No. 3 Starter Energizer to "ACCEL" with his other hand; this Starter Energizing Switch is located below the priming switch.

PRIME THE ENGINE CAREFULLY FOR STARTING—Open the No. 3 Throttle approximately ⅓; prime the No. 3 Engine. The primer switches are located below the Instrument Panel to the left of the Co-Pilot's Control Column and are numbered corresponding to the engines. To prime, press the switch intermittently. By doing this the fuel is driven into the engine intake in spurts and priming is much more effectively accomplished than by holding the switch "ON" for a fixed interval. From one to ten such "shots" are required depending upon the temperature of the engine and the outside air.

CRANK—With the engine properly primed and the starter energized the Co-Pilot now throws the meshing switch to "CRANK." There are two types of starters on the B-24: The earlier type where the Energizing Switch is held to "ACCEL" for thirty seconds then released before pressing the meshing switch, and the new type which is held to "ACCEL" for twelve seconds and then continued in the "ACCEL" position while the meshing switch is thrown to "CRANK." The latter type starter is a constant energizer and keeps the engine turning as long as the Energizing Switch is held to "ACCEL."

PLACE MIXTURE CONTROLS IN "AUTO-LEAN" WHEN ENGINE FIRES—After the engine definitely fires on the priming charge, throw the mixture controls from the "IDLE CUT-OFF" to the "AUTO-LEAN" position. "AUTO-LEAN" is used for starting, warm-up, and taxiing, as "AUTO-RICH" or "FULL-RICH" will cause the plugs to become fouled and the engine to load up because of the too rich mixture. When the mixture control is in "IDLE CUT-OFF" there is no flow of fuel to the carburetor jets. As soon as the mixture control lever is moved from "IDLE CUT-OFF" a valve is opened and fuel under the booster pump pressure flows to the carburetor jets. Therefore, if the engine does not start immediately, the mixture control must be returned to "IDLE CUT-OFF" to prevent flooding of the carburetor and the entire induction system. Do not exceed 1400 RPM in "AUTO-LEAN" on the ground as higher speeds, increased temperature and manifold pressure will cause detonation.

WARM-UP—With all engines running at 1000 RPM, warm-up is continued until the oil temperature gauge located on the Co-Pilot's Instrument Panel reaches 40° to 60° C. Each of the two temperature gauges has two hands, Engines Nos. 1 and 2 on the left gauge and Nos. 3 and 4 on the right gauge. The head temperature gauges likewise have two hands paired on each gauge in the same manner.

CHECK THE VACUUM PUMPS FOR PROPER OPERATION—There are two vacuum pumps, one on Engine No. 1 and one on Engine No. 2. While the engines are warming up both pumps are checked. The vacuum pump transfer or selector valve is located on the forward face of the bomb bay bulkhead on the left side, under the generator control panel. The valve instruction plate gives the source and the supply. It is a two position valve, No. 1 to De-Icer, No. 2 to Instruments, or vice versa. There is no "OFF" position. At 1000 RPM the vacuum gauge, located on the left of the Pilot's Instrument Panel, should read from 4 to 4½ inches.

CHECK THE PRESSURE IN THE ACCUMULATORS FOR BRAKE POWER—Check the brake pressure. The two gauges are located left of the Pilot's Control Column on the Instrument Panel. They register the pressure in each of the two accumulator systems. They should read between 850 and 1125 pounds. The top limit is 1050 pounds normally on the engine pump, but if the auxiliary hydraulic pump is operating it cuts out 75 pounds higher or at 1125.

CHECK THE DE-ICER AND ANTI-ICER OPERATION—The de-icer lever is located on the left of the Co-Pilot's Control Column under the switch panel. Throw it to the right for "ON" and check visually that the de-icer boots inflate and deflate. Turn the de-icer control "OFF," then check the anti-icer fluid flow to the propellers. The fluid motors are operated by a rheostat calibrated at Gallons Per Hour and located behind the mixture controls on the Instrument Panel. Check the flow of fluid at the propeller hub visually.

TAXI AWAY FROM THE LINE FOR FINAL RUN-UP—Unless there is some specific reason or local orders, the wise Pilot taxies the plane to the head of the runway for final run-up and preparation for take-off so that he can leave immediately thereafter and avoid the inevitable fouled engines. Then, too, planes warming up on the line are a definite hazard. Before taxiing, the wheel blocks are removed and when the last crew member enters the airplane, he pulls inboard on the auxiliary bomb door handle located immediately forward of the bomb bay bulkhead on the right side under the flight deck. This closes the bomb doors. Upon pressing the brake pedals the parking lever automatically disengages and releases the parking brakes.

PROPELLER MUST BE SET IN HIGH RPM FOR MAXIMUM TAKE-OFF POWER—Before taxiing away from the line, the Pilot again checks the propellers for High RPM by throwing the propeller switch bar to "INC" RPM (Increase RPM). When the propeller controls are in the full extreme positions, either High or Low RPM, the four Governor Limit Indicator Lights on the center of the Instrument Panel are illuminated.

TAXIING—Too often a carelessly executed maneuver—The maneuver of trying to taxi from the line should be executed in a broad sweep, using the power of the outboard engines and not the brakes to make the turn. Sharp turns should be definitely avoided. They exert severe strain on the landing gear and on the tires. It is a simple matter to turn and steer the B-24 with the outboard engines. The nose wheel is limited to 45° either side of the centerline. Turns which require an angularity greater than this bring the nose mechanism up against the stops and can cause failure. The strains, exerted by such a maneuver, particularly if done violently, are extremely severe. Remember—the B-24 is a heavy airplane. Sharp turns do not allow the inboard wheel to follow a track. As a result, due to the wide contact surface, a quantity of rubber—which can not be ignored, is scraped from the inner tire. Severe stresses are imposed on the entire landing gear assembly. The Engineers, in designing an airplane, expect the Pilots to exercise a reasonable amount of good judgment.

Always taxi at low speed and maintain direction with the outboard engines and rudders. When it is necessary to use the brakes, apply them gently and your airplane will always be under control.

THE WARM-UP—On arriving at the head of the runway or the designated warm-up area, stop the airplane and set the parking brakes.

EXERCISE THE PROPELLERS TO CHECK THEIR FULL RANGE OPERATION—With the throttles set for 1000 RPM on all engines exercise the propellers throughout their entire range. The Propeller Indicator Lights will come "ON" when the propellers reach the limit of travel. Exercise all the way to Low RPM high pitch and return to High RPM low pitch. Leave them in this position for the run-up and take-off.

ADJUST THE CONTROL TABS FOR TAKE-OFF—The control tabs are adjusted during the run-up for the best take-off positions; the elevator tabs are set for 1° tail heavy; aileron tab is set in neutral; rudder tabs are set for 1° to 2° right rudder. These are the settings for normal conditions and loading. At this time it is wise to check the Automatic Flight Control and De-Icer again as well as controls for freedom of movement. These checks become automatic with experience as they affect the control and performance of the airplane.

"AUTO-RICH" MIXTURE POSITION IS ALWAYS USED FOR TAKE-OFF AND FOR FULL POWER REQUIREMENTS—Preparatory to the run-up the mixture controls are moved from "AUTO-LEAN" and set in the "AUTO-RICH" position. In "AUTO-LEAN" the engines will detonate under full power. Completely detonating engines will stop completely and suddenly.

CHECK THE FUEL PRESSURE AND THEN TURN "ON" THE BOOSTER PUMPS—Check the fuel pressure with the booster pumps "OFF"—The normal reading should be from 14 to 16 pounds and then turn the booster pumps "ON" for run-up and take-off. The booster pump switches are the top row of switches under the Instrument Panel to the left of the Co-Pilot's Control Column. The engines are run up with at least 105° C. head temperature but not over 205° C. Engines are run-up one at a time.

RESTRICTED

CHECK BOTH MAGNETOS AT 2000 RPM—During the run-up when the engine reaches 2000 RPM both switches are checked. A maximum of 75 RPM drop on either magneto is allowable. After this check, open the throttles wide against the stop. The engine supercharger will give approximately 36 to 37 inches manifold pressure at sea level.

SET THE SUPERCHARGER FOR THE TAKE-OFF REQUIREMENTS—With the throttle wide open move the supercharger control forward slowly to "ON". The manifold pressure increases from the engine supercharger boost to the final sea level take-off setting of 49 inches. The turbo-supercharger increases the boost 12 inches by creating a pressure, or ram, on the inlet air to the carburetor. The turbo control stops are set at the factory for a 49 inch sea level reading. This same run-up procedure is carried out for all four engines.

EXTEND THE FLAPS FOR TAKE-OFF ON NO. 3 RUN-UP—On the run-up of No. 3 Engine we take advantage of the hydraulic power supply, and extend the wing flaps to the take-off setting: either 10° or 20° depending on take-off conditions. The Fowler flaps are set to 10° for the best average take-off. This setting gives maximum performance in case of engine failure. Flaps extended 20° is the setting for maximum take-off lift. For normal operation, however, the 10° setting is recommended. To stop the flaps in any position before "FULL DOWN" return the operating lever to neutral manually. In the extreme positions the lever returns automatically to neutral. After run-up it is better to maintain 800 to 1000 RPM idling speed so that the engines will not foul up. Low idling speeds cause sooty and malfunctioning spark plugs.

TURN GENERATORS "ON" AFTER RUN-UP—The four generators are cut in at the panel, located outboard of the fuel gauges on the left forward face of the bomb bay bulkhead. As each switch is turned to "ON" increase the revolutions of the same numbered engine slightly to check the charge. The generators are left "OFF" during warm-up. If the generator switches are turned "ON" the voltage regulator points become badly mutilated due to the excess vibration during warm-up.

CLOSE THE COWL FLAPS FOR TAKE-OFF—The cowl flaps are closed to "⅓ OPEN" for take-off. Normally this gives ample cooling with a minimum of resistance for take-off. For normal temperature conditions, less than "⅓ OPEN" will not provide sufficient cooling. It is a point to remember that on an average the cowl flaps reduce the airplane speed eight-tenths of a mile per hour at cruising speed for each degree of cowl flap opening. With the cowl flaps open more than ⅓, tail buffeting results. A wider opening, therefore, is not recommended for take-off nor in flight unless absolutely necessary due to extreme heat conditions.

LANDING LEVER PLACED IN "DOWN" FOR CHECK—After the run-up, move the landing gear lever to the "DOWN" position again and check the kick-out pressure. It should read from 825 to 875 p.s.i. on the Main Hydraulic Gauge. This is located above the Pilot's Control Column. Glance at the instrument for a last look. Check the crew aboard and be sure that the Nose Wheel Compartment is clear. Now all is in readiness for the take-off run.

THE TAKE-OFF—Release the brakes, swing into the wind and open the throttle slowly and evenly against the stops. Have the Co-Pilot hold them in this position so there will be no chance of their creeping closed. The throttle frictions are usually set lightly for take-off and landing to permit free movement if necessary. During the take-off run the Co-Pilot must check the supercharger pressure carefully and make any adjustment to hold the supercharger pressures equalized at 49 inches. As the plane accelerates the Pilot should apply a gentle back pressure on the controls to assist in lifting the plane on the gear. The plane, with a moderate load, leaves the ground easily at 110 MPH. This take-off speed increases up to 130 MPH for a plane with a full load. After leaving the ground the nose of the plane should be held down and the take-off course maintained until the indicated airspeed reaches 135 MPH. At this speed full control is available in the event of an outboard engine failure, under average conditions.

RAISE LANDING GEAR—As soon as the airplane is well clear of the ground and definitely air borne the landing gear is raised. If the landing gear fails to retract immediately the cause is probably air in the system, which can be eliminated by working the operating lever through sufficient "UP" and "DOWN" cycles to bleed it off.

REDUCE THE POWER TO SAVE ENGINES—After take-off reduce the manifold pressure to 45.5 inches (the maximum allowable continuous power rating for one hour) by retarding the turbo control lever.

REDUCE PROPELLER REVOLUTIONS TO 2550 AND SYNCHRONIZE PROPELLERS—Reduce the revolutions to 2550 with the propeller control switches.

RAISE THE FLAPS—After the landing gear lever has returned to neutral the flaps may be raised. Do not attempt to operate the landing gear and the flaps simultaneously; with the open center system the valve nearest the engine pump cuts off all other units.

Airspeed of 155 MPH must not be exceeded with the flaps extended.

FUEL BOOSTER PUMPS "OFF"—The fuel booster pumps are turned "OFF" as their auxiliary pressure is not again needed until the fuel pressure drops 2 pounds due to altitude or until landing.

ADJUST COWL FLAPS—Cowl flaps are adjusted as necessary to control engine head temperatures not to exceed 260° C. in climb (or under maximum power) with Mixture Controls in "AUTO-RICH."

The maximum head temperature allowable for cruising in "AUTO-LEAN" is 232° C. Check the engine instruments:

Oil pressure 75 to 80 pounds;

Fuel pressure 14 to 16 pounds;

Maximum oil temperature for "AUTO-LEAN" cruising 75° C.

For full rate of power in "AUTO-RICH" a maximum oil temperature of 85° C. is permitted. For military power for five minutes 100° C. is the maximum allowable.

To reduce the drag the cowl flaps should be kept as nearly fully closed as possible. High airspeed cools better than open cowl flaps. The use of cowl flaps reduces lift as well as increases drag; therefore, use minimum opening which will maintain proper cooling. Do not allow head temperature to exceed 260° for full power or high speed in "AUTO-RICH" or 232° C. for continuous operation in "AUTO-LEAN."

TURN THE AUXILIARY HYDRAULIC MOTOR "OFF" ON EXTENDED FLIGHTS—The auxiliary hydraulic electric motor is turned "OFF" when other than a purely local flight is being made. The hydraulic pump on No. 3 Engine is capable of furnishing all the necessary power for flight operations. The auxiliary pressure switch has maximum and minimum setting such that the unloading valve will not function and the electric pump when "ON" will supply all of the accumulator charge.

LANDING—As the plane approaches the field and enters the pattern the Pilot retards the throttles and reduces the speed to 155 MPH.

NOTIFY THE CREW SO THEY CAN PERFORM THEIR DUTIES—As speed is being reduced the Pilot notifies the crew that the airplane is coming in for a landing and receives a report that the Nose Wheel Compartment is clear of the crew and that all is in readiness for a landing.

TURN "ON" HYDRAULIC ELECTRIC SWITCH FOR AUXILIARY POWER—The auxiliary hydraulic switch is turned "ON" as the full supply of main hydraulic power will not be available when No. 3 Engine is throttled.

CHECK ACCUMULATOR PRESSURE TO BE SURE OF BRAKING POWER—Check the accumulators for proper pressure. The gauge is located on the left of the Pilot's Control Column and should read between 850 and 1125 p.s.i.

TURN "OFF" A.F.C. OR "AUTOMATIC PILOT"—Make sure that the "Automatic Flight Control" or "Pilot" is turned "OFF." Landing would be hazardous attempting to overpower the automatic controls.

CLOSE THE COWL FLAPS TO REDUCE DRAG AND RETARD ENGINE COOLING—Cowl flaps are closed on the approach to prevent rapid engine cooling in the glide and to cut down head resistance in the event landing is refused. Open cowl flaps also lower the lift of the wing surface directly behind them which is a considerable area.

PLACE THE MIXTURE CONTROLS IN "AUTOMATIC RICH"—The mixture controls are placed in the "AUTO-RICH" position in the event full power might be needed. (Full power is available only with the mixture controls in "AUTO-RICH.")

INTERCOOLER SHUTTERS MUST BE "OPEN"—Intercooler shutters are checked for "OPEN" unless they are needed because of carburetor icing; in which case the head temperatures must be watched carefully and the Co-Pilot on the alert to open them immediately.

TURN "ON" THE BOOSTER PUMPS—Booster pumps are turned "ON" to insure positive flow of fuel to the engine pumps.

THE DE-ICERS MUST BE "OFF"—The De-Icers must be turned "OFF"—Be sure to check this. When the De-Icers are operating the inflated shoes act as spoilers as the wing approaches the stall and change the landing behavior of the airplane.

CHECK FOR LANDING KICK-OUT PRESSURE WITH LANDING GEAR LEVER IN "UP" POSITION—Move the landing gear lever to the "UP" position to check kick-out pressure; which should be from 1050 to 1100 p.s.i.

LOWER LANDING GEAR WHEN SPEED IS REDUCED TO 155 MPH—Move the landing gear lever, located on the left side of the pedestal, to the rear and downward to the "DOWN" position. Lever will return to neutral when the gear is down. As the gear is lowering check this sequence of operation: The hydraulic pressure on the main gauge to the left of the Pilot's Control Column builds up suddenly and then drops; the warning light in front of the Pilot on the Instrument Panel turns "ON"; landing gear control handle returns to neutral. The warning horn sounds when the throttle is closed unless the gear is latched properly in the "DOWN" position. The return of the handle to neutral does not mean that the latches are engaged. A surge as the gear bottoms could cause a premature kickout.

INSPECT ALL LANDING GEAR LATCHES VISUALLY—A crew member must check the gear latches to be absolutely certain they are engaged. The nose gear latches may be inspected from the Nose Wheel Compartment. Each of the main landing gear latches can be seen from the rear window on each side. They are painted a bright yellow for immediate identification. The main landing gear latches can not be seen with the flaps extended.

LOWER THE WING FLAPS HALF-DOWN FIRST—After the landing gear lever has kicked back to neutral and the gear has been checked; with the speed still reduced to 155 MPH, enter the landing lane and extend the wing flaps 20° by moving the flap lever on the right of the pedestal to the rear. When the flap indicator reads 20° return the flap control lever to neutral manually. This stops the flaps in the 20° "DOWN" position. The flap lever only returns automatically from the extreme positions "UP" and "DOWN." Half-down flap is recommended for the beginning of the approach. With the flaps in this position the lift and drag are both increased and the attitude of the airplane affords a greater angle of vision during landing. With the flaps partially or completely extended the airplane is fully manueverable but not so responsive.

TURBO CONTROLS ARE TURNED "OFF"—Turbo controls are turned "OFF" normally. When landing at altitude they are left in a position to furnish the required manifold pressure. Handle the throttles carefully. With the turbo controls "ON" a backfire may blow off the exhaust manifold as the tail gate is closed and the turbo outlet restricts free exhaust.

PROPELLERS HIGH RPM—Place the propeller control switches at "INC" RPM. This throws the propellers in low pitch High RPM so that maximum power will be available in case landing is refused.

FLAPS "FULL DOWN"—When approaching the boundary of the landing field the flap lever is placed in the "DOWN" position and flaps fully extended. The lever will return automatically to neutral and the indicator will show 40° extended flap. When the flaps are extended fully always allow sufficient interval of time before the final level-off, or flare, for the airplane to settle into its new attitude to avoid confusing the Pilot at the last minute before the flare for landing.

CHECK THE LANDING GEAR "DOWN" AGAIN—Move the landing gear lever again to the "DOWN" position for final check.

CLOSE THE THROTTLES AND LAND—When the airplane reaches the proper position over the runway, with the throttles closed, begin the flare with ample altitude for control response. Remember, the B-24D Airplane has great momentum due to its weight and resists sudden change of direction. Adjust the elevator tabs to assist in the landing and hold the airplane off the ground as long as possible. The best landing position is the conventional one for airplanes not equipped with tricycle gear. Never land in a position which will allow the nose wheel to make contact first. A three-wheel landing should be made only when brake application is necessary immediately upon touching the ground.

WARSHIPS DVD SERIES

WARSHIPS: CARRIER MISHAPS

AIRCRAFT CARRIER MISHAPS
SAFETY AND TRAINING FILMS

-PERISCOPEFILM.COM-

Now Available on DVD!

Warships DVD Series

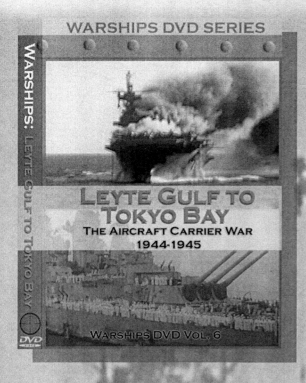

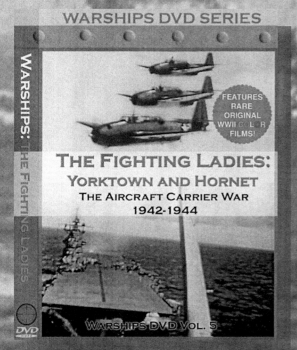

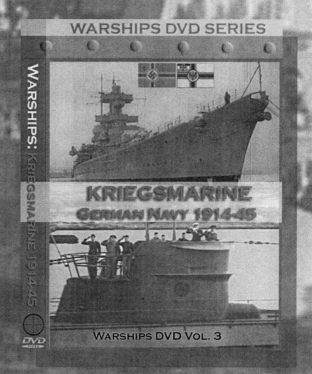

Now Available!

Aircraft At War DVD Series

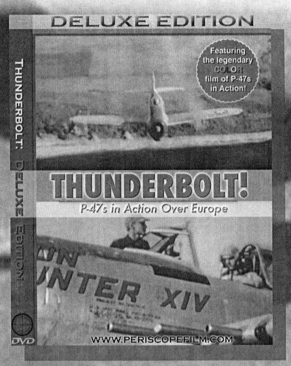

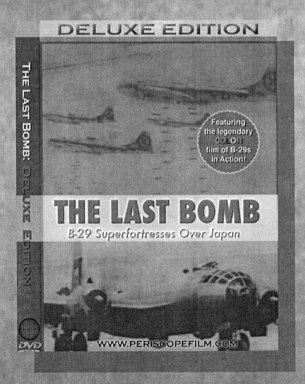

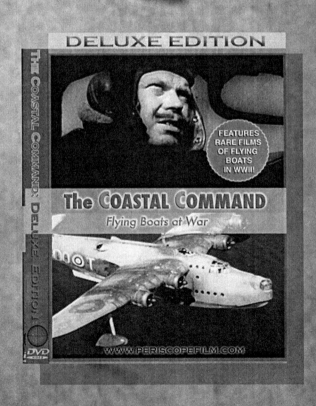

Now Available!

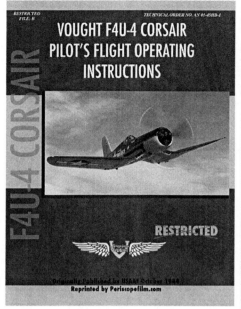
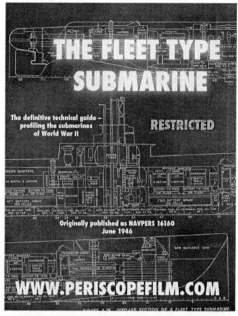
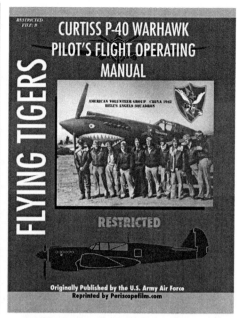
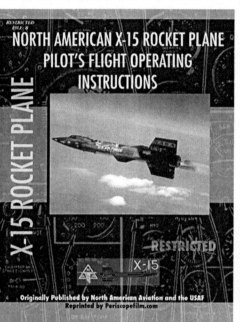
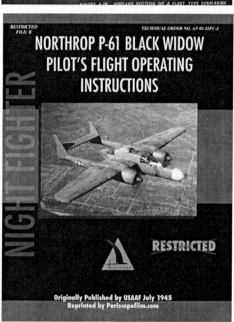
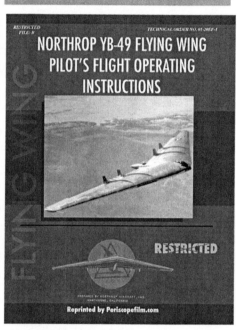

ALSO NOW AVAILABLE FROM PERISCOPEFILM.COM

©2006-2009 Periscope Film LLC
All Rights Reserved
ISBN #978-1-4116-1321-8

LaVergne, TN USA
03 February 2011
214920LV00002B/2/P